ARCHIVES OF DATA SCIENCE

SERIES A

Special Issue: Selected Papers of the 3rd German-Polish Symposium on Data Analysis and Applications

Volume 1 | Number 1 | May 2016

ARCHIVES OF DATA SCIENCE

SERIES A

Special Issue: Selected Papers of the 3rd German-Polish Symposium on Data Analysis and Applications

Volume 1 | Number 1 | May 2016

Andreas Geyer-Schulz Institute of Information Systems and Marketing, Information Services and Electronic Markets, Karlsruhe Institute of Technology (KIT), Karlsruhe, E-mail: andreas.geyer-schulz@kit.edu

Józef Pociecha Department of Statistics, Cracow University, Kraków E-mail: jozef.pociecha@uek.krakow.pl

KIT Scientific Publishing

Archives of Data Science, Series A publishes papers of short to medium length (approximately 8 - 20 pages) in the emerging field of data science. It covers regular research articles from the field of Data Science and special issues on conferences, workshops and joint activities of the German Classification Society/Gesellschaft für Klassifikation e.V. (GfKl (http://www.gfkl.org/en/)) and its cooperating partners and organisations.

Impressum

KIT Scientific Publishing

Karlsruher Institut für Technologie (KIT)
KIT Scientific Publishing
Straße am Forum 2
D-76131 Karlsruhe

KIT Scientific Publishing is a registered trademark of Karlsruhe Institute of Technology. Reprint using the book cover is not allowed.

www.ksp.kit.edu

Print on Demand 2017 – Gedruckt auf FSC-zertifiziertem Papier

ISSN 2363-9881
ISBN 978-3-7315-0581-5
DOI 10.5445/KSP/1000058747

Contents

Editorial

Andreas Geyer-Schulz and Józef Pociecha

The first volume of Archives of Data Science, Series A is a special issue of a selection of contributions which have been originally presented at the *3rd Bilateral German-Polish Symposium on Data Analysis and Its Applications* (GPSDAA 2013). The GPSDAA is a joint workshop of the Gesellschaft für Klassifikation e.V. (GfKl) and the Sekcja Klasyfikacji i Analizy Danych (SKAD) and its third incarnation was hosted by Hermann Locarek-Junge at the TU Dresden from September 26th-28th, 2013.

The workshop contained 22 presentations of joint work in progress in data science the following researchers:

Werner Adler, Daniel Baier, Fabian Ball, Tomasz Bartłomowicz, Andrzej Bąk, Andreas Beyer, Hans-Hermann Bock, Robert Busa-Fekete, Sabina Denkowska, Józef Dziechciarz, Manuel Ferreira, Andreas Geyer-Schulz, Thomas Górecki, Swetlana Herbrandt, Eyke Hüllermeier, Hans A. Kestler, Zardad Khan, Anna Król, Michael Kuhn, Katarzyna Kuziak, Berthold Lausen, Ludwig Lausser, Paweł Lula, Hans-Joachim Mucha, Jan Mutl, Jan W. Owsiński, Barbara

Andreas Geyer-Schulz
Information Services and Electronic Markets, Karlsruhe Institute of Technology (KIT)
Kaiserstraße 12, D-76131 Karlsruhe, Germany
✉ Andreas.Geyer-Schulz@kit.edu

Józef Pociecha
Department of Statistics, Cracow University of Economics,
Rakowicka 27, 31-510 Kraków, Poland

✉ Jozef.Pociecha@uek.krakow.pl

Archives of Data Science, Series A
KIT Scientific Publishing
Vol. 1, No. 1, S. 1–2, 2016

DOI 10.5445/KSP/1000058747/00
ISSN 2363-9881

Pawełek, Marcin Pełka, Krzysztof Piontek, Józef Pociecha, Sergej Potapov, Nils Raabe, Christian Rautert, Aneta Rybicka, Adam Sagan, Stefanie Schreiber, Leopold Sögner, Andrzej Sokołowski, Christoph Stadtfeld, Marta Targaszewska, Claus Weihs, Justyna Wilk, and Waldemar Wołyński.

The nine papers in this volume are all revised and considerably extended versions of the presentations given at the GPSDAA 2013. We thank all authors for the considerable additional effort and time they have put into the improvement of their papers and, last but not least, for their patience with the delays in the publication process.

On Friday, September 27th, 2013 a roundtable on the *Future of Scientific Publishing* was held which led to vivid discussions among the workshop participants. In a way these discussions were also crucial for the creation of the Archives of Data Science, Series A.

This issue is structured by topics:

Clustering: Hans-Hermann Bock contributes a survey paper on probabilistic two-way clustering and Hans-Joachim Mucha a paper on the problem of assessing cluster stability.

Machine Learning: Michael Bräuning and Eyke Hüllermeier present new algorithms for learning conditional lexicographic preference trees and Justyna Wilk shows an application of classification trees to demographic research.

Conjoint Analysis: Daniel Baier, Marcin Pełka, Aneta Rybicka and Stefanie Schreiber compare traditional and choice-base conjoint analysis models when both models are estimated by hierarchical Bayes algorithms and Tomasz Bartłomowicz, Andrzej Bąk present the MaxDiff R-package.

Applications: The last group of papers covers three different areas of applications of data science methods:

- Józef Dziechciarz, Anna Król, Marta Targaszewska present a variety of approaches of measuring the effectiveness of tertiary education.
- Claus Weihs, Swetlana Herbrandt, Nils Raabe, Manuel Ferreira, Christian Rautert combine laboratory experiments, statistical modelling techniques and finite element methods in their approach to simulate grinding processes with diamond tools.
- Ludwig Lausser, Florian Schmid, Matthias Platzer, Mikko J. Sillanpää, and Hans A. Kestler apply classifiers to develop diagnostic tools for gene expression data which bridge the gap to higher-level explanations such as molecular signaling pathways.

Probabilistic Two-way Clustering Approaches with Emphasis on the Maximum Interaction Criterion

Hans-Hermann Bock

Abstract We consider the problem of simultaneously and optimally clustering the rows and columns of a real-valued $I \times J$ data matrix $X = (x_{ij})$ by corresponding row and columns partitions $\mathscr{A} = (A_1, ..., A_m)$ and $\mathscr{B} = (B_1, ..., B_n)$, with given m and n. We emphasize the need to base the clustering method on a probabilistic model for the data and then to use standard methods from statistics (e.g., maximum likelihood, divergence) to characterize optimum two-way classifications. We survey some clustering criteria and algorithms proposed in the literature for various data types. Special emphasis is given to the maximum interaction clustering criterion proposed by the author in 1980. It can be shown that it results as the maximum likelihood clustering method under a two-way ANOVA model (with individual main effects, but cluster-specific interactions). After a simple data transformation (double-centering) well-known two-way SSQ clustering algorithms can directly be used for maximization.

Hans-Hermann Bock
Institute of Statistics, RWTH Aachen University, Vorder-Winterbach 36, 77794 Lautenbach, Germany, ✉ bock@stochastik.rwth-aachen.de

ARCHIVES OF DATA SCIENCE, SERIES A
KIT SCIENTIFIC PUBLISHING
Vol. 1, No. 1, S. 3–20, 2016

DOI 10.5445/KSP/1000058747/01
ISSN 2363-9881

1 Two-way clustering problems

Two-way clustering means clustering, simultaneously, the rows and columns of a data matrix $X = (x_{ij})_{I\times J}$. Synonymns are bi-clustering, co-clustering, or block clustering. In practice, two-way clustering problems occur, e.g.,

- in microbiology (microarray measurements for I genes and J different times, situations, or tissues); see, e.g., Martella et al (2008), Cheng and Church (2000), Madeira and Oliveira (2004), Martella et al (2011), Martella and Vichi (2012), Turner et al (2005)
- in marketing (purchase data for I consumers described by J social characteristics); see, e.g., Baier et al (1997), Arabie et al (1988)
- in documentation (I documents or e-mails described by presence/absence of J keywords); see, e.g., Dhillon et al (2003), Banerjee et al (2007), Li and Zha (2006), Cho et al (2004), Cho and Dhillon (2008).

Many two-way clustering methods have been proposed since the beginning of clustering activities in the 1970s (recent surveys were given by Van Mechelen et al, 2004; Madeira and Oliveira, 2004; Charrad and Ben Ahmed, 2011; Vichi, 2012; Govaert and Nadif, 2013), but the possibility to record automatically huge sets of data in various application fields has meanwhile increased the importance of two-way clustering for an adequate and informative analysis of data.

In this paper we consider a real-valued data matrix $X = (x_{ij})_{I\times J}$ with I rows, J columns and try to find an m-partition $\mathscr{A} = (A_1,...,A_m)$ of the row set $\mathscr{I} = \{1,...,I\}$ with m classes, and an n-partition $\mathscr{B} = (B_1,...,B_n)$ of the column set $\mathscr{C} = \{1,...,J\}$ with n classes, such that the joint $m\cdot n$-partition $\mathscr{A}\times\mathscr{B} = \{A_r \times B_s | r = 1,...,m, s = 1,...,n\}$ of the set of pairs $\{(i, j)|i \in \mathscr{I}, j \in \mathscr{J}\}$ (cells of the matrix X) together with a suitable parametric characterization of the classes fits, approximates or reproduces optimally the hidden row by column structure (if any) in the given data matrix X. Obviously, such a formulation requires the specification of some "structure" that should be reconstructed from the data, and some optimality criterion that should be optimized. The multitude of proposed two-way clustering algorithms can be largely explained by the great number of choices for "structure" and "optimality".

We emphasize here the probabilistic approach where "structure" is described by a parametric and block-specific probability distribution for the data X_{ij}. Then, generally, the parameter estimates as well as the bi-clustering $(\mathscr{A}, \mathscr{B})$ are obtained by the maximum-likelihood (m.l.) approach. Thereby, the choice of a distributional model is highly dependent on the way in which the data were obtained and on their interpretation as measurement values, associations, frequencies, indicators, etc. In this respect we will consider

- association-type data for a two-mode data matrix (Sect. 2
- measurement-type values x_{ij} with categorical factor levels i, j (Sect. 3)
- frequency-type values N_{ij} with factor levels i, j (contingency table; Sect. 4)
- object by variable measurements x_{ij} (classical data matrix; Sect. 5)

and provide some exemplary probabilistic clustering approaches. For binary variables we refer, e.g., to Govaert and Nadif (2005); Li (2005); Govaert and Nadif (2007, 2008, 2013) and Nadif and Govaert (2010).

Note that we will not comment here on the choice of the numbers m, n of classes (see, e.g., Schepers et al, 2008) and will present only the so-called "fixed-partition" or "classification likelihood" approaches (see, e.g., Bock, 1996a,b). Alternatively, probabilistic clustering approaches can also be formulated in terms of mixture models ('random-partition" approach) resulting in EM-type algorithms and fuzzy bi-partitions in the form of posterior distributions (see, e.g., Govaert, 1995; Govaert and Nadif, 2005, 2003, 2008, 2010; Bocci et al, 2006; Li and Zha, 2006; Martella et al, 2008, 2011). Other approaches use row- and column-wise hierarchical clusterings or try to cover the set of IJ matrix cells with suitably weighted, possibly overlapping "homogenous blocks" $A \times B$ such as *plaid methods* (described by Lazzeroni and Owen, 2002; Turner et al, 2005) or *additive clustering* (as in Shepard and Arabie, 1979; Mirkin et al, 1995; Wilderjans et al, 2013). See also the articles on multi-mode clustering in the Special Issue on "Statistical learning methods including dimension reduction" of the journal "Computational Statistics and Data Analysis" (vol. 52, 2007, edited by H.-H. Bock and M. Vichi).

2 Clustering for association-type data

In this section we suppose that the data x_{ij} represent association values that measure how "close", "associated", or "interrelated" row i is to column j. Also we assume a two-mode case, i.e., rows and columns refer to different sets (such as customers and products, genes and time points, respectively). In this case a classical two-way clustering criterion is provided by the SSQ:

$$g(\mathscr{A},\mathscr{B},\mu) := \sum_{r=1}^{m}\sum_{s=1}^{n}\sum_{i\in A_r}\sum_{j\in B_s} ||x_{ij}-\mu_{rs}||^2 \to min_{\mathscr{A},\mathscr{B},\mu} \qquad (1)$$

where $\mu_{rs} \in R$ is a block-specific prototype value and μ the set of these values[1] (Bock, 1980). This criterion amounts to approximating the given data matrix X by an "ideal" block-matrix $\widetilde{X}_{I\times J}$ with the same value μ_{rs} in all cells of a block (bicluster) $A_r \times B_s$ (for all r,s). Given that partial minimization with respect to μ leads to the average values $\hat{\mu}_{rs} = \bar{x}_{A_r\times B_s}$ in the blocks $A_r \times B_s$ of X, the criterion (1) is equivalent to the following *SSQ clustering criterion*:

$$Q_{min}(\mathscr{A},\mathscr{B};X) := \sum_{r=1}^{m}\sum_{s=1}^{n}\sum_{i\in A_r}\sum_{j\in B_s} ||x_{ij}-\bar{x}_{A_r\times B_s}||^2 \to \min_{\mathscr{A},\mathscr{B}} \qquad (2)$$

and to

$$k(\mathscr{A},\mathscr{B};X) := \sum_{r=1}^{m}\sum_{s=1}^{n} |A_r|\cdot|B_s|\cdot||\bar{x}_{A_r\times B_s}||^2 \to \max_{\mathscr{A},\mathscr{B}}. \qquad (3)$$

In order to optimize these clustering criteria many algorithms (e.g., double k-means) have been proposed; see, e.g., Bock (1980); Gaul and Schader (1996); Baier et al (1997); Hansohm (2002); Vichi (2001); Castillo and Trejos (2002); Cho et al (2004); Cho and Dhillon (2008); Rocci and Vichi (2008); Van Rosmalen et al (2009); Schepers and Hofmans (2009); Martella and Vichi (2012)

3 Clustering for factorial designs

In this section we consider the case where all data values x_{ij} are measurements of the same target variable which, however, depends on two categorical factors U (rows) and V (columns) with categories in $\mathscr{I} = \{1,...,I\}$ and $\mathscr{J} = \{1,...,J\}$, respectively. For example, in a diet experiment with many persons, U might be the initial BMI (discretized body mass index, $I = 30$, say) of a person, V the type of diet that this person applies (with $J = 15$ types, say), and x_{ij} the average loss of weight after a four-weeks diet for all persons with $U = i$ and $V = j$. Assuming a complete factorial design (i.e., observations were made for

[1] $||x||$ means the absolute value $|x|$ for $x \in R^1$ and the Euclidean norm for multivariate data (see Remark 2). For a set A, $|A|$ means the number of elements of A.

all IJ combinations $(i,j) \in \mathscr{I} \times \mathscr{J}$) the clustering problem consists in finding (a given number $m = 6$, say, of) BMI classes $A_1, ..., A_m$ and (a given number $n = 4$, say, of) diet classes $B_1, ..., B_n$ that best describe the data. In this way, the large number of categories can be reduced to a smaller and handy number of category classes or "types".

Classical statistics analyzes such two-way configurations by ANOVA models with random variables X_{ij} that are additively obtained from a total mean, row and column main effects, interaction terms, and normal errors. In the clustering framework we consider two such models: one with individual main effects, and one with class-specific main effects. It appears that only the first one provides new insights while the second one falls back to the criterion (2).

3.1 ANOVA clustering model with individual main effects

Here we assume that the existence of a hidden bi-clustering is exclusively caused by block-specific interaction terms while main effects do not contribute to the clustering aspect. In the framework of ANOVA this amounts to suppose that X_{ij} are given, for a fixed bi-partition $(\mathscr{A}, \mathscr{B})$, by the additive composition:

$$X_{ij} = c + a_i + b_j + \gamma_{rs} + e_{ij} \qquad i \in A_r, j \in B_s, r = 1, ..., m, s = 1, ..., n. \quad (4)$$

Here c is a fixed mean value, a_i the *individual* main effect of category i of U, b_j the *individual* main effect of category j of V, and γ_{rs} the *class-specific* interaction effect; the latter one is the same for all pairs (i,j) in the bicluster $A_r \times B_s$. The e_{ij} are independent random error terms with $e_{ij} \sim \mathscr{N}(0, \sigma^2)$ where we consider σ^2 to be known here (but see Remark 1). In order to attain identifiability of parameters, the following zero-means normalization is introduced:

$$\begin{aligned} &\bar{a}_\bullet := \textstyle\sum_{i=1}^{I} a_i/I = 0, && \bar{b}_\bullet := \textstyle\sum_{j=1}^{J} b_j/J = 0, \\ &\bar{\gamma}_{\bullet,s} := \textstyle\sum_{r=1}^{m} |A_r| \cdot \gamma_{rs}/I = 0, && \bar{\gamma}_{r,\bullet} := \textstyle\sum_{s=1}^{n} |B_s| \cdot \gamma_{rs}/J = 0 \qquad \text{for all } r, s. \end{aligned}$$

For estimating the unknown parameters c, a_i, b_j, γ_{rs} and the unknown $(\mathscr{A}, \mathscr{B})$ we use the m.l. approach. Due to the normality assumptions this amounts to minimizing the SSQ:

$$\widetilde{Q}(c,a,b,\gamma,\mathscr{A},\mathscr{B}) := \sum_{r=1}^{m} \sum_{s=1}^{n} \sum_{i \in A_r} \sum_{j \in B_s} ||x_{ij} - c - a_i - b_j - \gamma_{rs}||^2 \to \min_{c,a,b,\gamma,\mathscr{A},\mathscr{B}} \quad (5)$$

After some algebraic manipulations (or using derivatives) we obtain, for a fixed bi-partition $(\mathscr{A},\mathscr{B})$, the following m.l. estimates:

$\hat{c} = \bar{x}_{\bullet,\bullet}$ overall mean

$\hat{a}_i = \bar{x}_{i,\bullet} - \bar{x}_{\bullet,\bullet}$ and $\hat{b}_j = \bar{x}_{\bullet,j} - \bar{x}_{\bullet,\bullet}$ individual main effects

$\hat{\gamma}_{rs} = \bar{x}_{A_r \times B_s} - \bar{x}_{A_r,\bullet} - \bar{x}_{\bullet,B_s} + \bar{x}_{\bullet,\bullet}$ class-specific interaction effects.

Inserting these estimates into (5) yields the clustering criterion:

$$\widetilde{Q}_{min}(\mathscr{A},\mathscr{B}) := \sum_{r=1}^{m}\sum_{s=1}^{n}\sum_{(i,j)\in A_r\times B_s}(x_{ij} - \hat{\mu} - \hat{a}_i - \hat{b}_j - \hat{\gamma}_{rs})^2 \quad \to \quad \min_{\mathscr{A},\mathscr{B}} \tag{6}$$

that can be shown, by algebraic transformations (see Bock, 1980; Schepers et al, 2013), to be equivalent to the following *maximum interaction clustering criterion*:

$$G(\mathscr{A},\mathscr{B};X) := \sum_{r=1}^{m}\sum_{s=1}^{n}|A_r|\cdot|B_s|\cdot|\hat{\gamma}_{rs}^{(X)}|^2 \tag{7}$$

$$= \sum_{r=1}^{m}\sum_{s=1}^{n}|A_r|\cdot|B_s|\cdot(\bar{x}_{A_r\times B_s} - \bar{x}_{A_r,\bullet} - \bar{x}_{\bullet,B_s} + \bar{x}_{\bullet,\bullet})^2 \to \max_{\mathscr{A},\mathscr{B}}$$

where we have flagged $\hat{\gamma}_{rs}^{(X)}$ by the superscript X in order to emphasize the corresponding data matrix X.

This clustering criterion was proposed by Bock (1980) on empirical grounds. The previous argumentation shows that it derives from the probabilistic factorial ANOVA approach (4). In Sect. 4 we will show that its minimization can be easily performed by the algorithms that were developed for the SSQ cluster criterion (2); so no specific algorithms have to be developed for (7).

Remark 1: It can easily be shown that the criterion (7) results as the m.l. clustering criterion also in the case of an unknown variance σ^2.

Remark 2: In case of vector-valued variables X_{ij} and observations $x_{ij} \in R^p$ the ANOVA model (4) must be formulated with p-dimensional effects c, a_i, b_j, γ_{rs} and $e_{ij} \sim \mathscr{N}_p(0, I_p)$. For this p-dimensional version the m.l. clustering approach yields the same clustering criteria as before (in particular, the maximum interaction criterion (7)) where $||...||$ now is the Euclidean norm in R^p.

3.2 ANOVA clustering model with class-specific main effects

We may wonder what happens if we assume that in the ANOVA model (4) not only the interactions, but also the main effects are class-specific. This amounts to the additive model

$$X_{ij} = \mu_{rs} + e_{ij} = c + \alpha_r + \beta_s + \gamma_{rs} + e_{ij} \quad i \in A_r, j \in B_s, r = 1, ..., m, s = 1, ..., n \tag{8}$$

with class-specific "block prototypes" $\mu_{rs} = c + \alpha_r + \beta_s + \gamma_{rs}$, typically with a zero-mean standardization for the effects $\alpha_r, \beta_s, \gamma_{rs}$. Note that for given $\{\mu_{rs}\}$ the standardized effects are uniquely determined by $c = \bar{\mu}_{\bullet,\bullet}$, $\alpha_r := \bar{\mu}_{A_r,\bullet} - \bar{\mu}_{\bullet,\bullet}$, $\beta_s = \bar{\mu}_{\bullet,B_s} - \bar{\mu}_{\bullet,\bullet}$ and $\gamma_{rs} = \bar{\mu}_{A_r,B_s} - \bar{\mu}_{A_r,\bullet} - \bar{\mu}_{\bullet,B_s} + \bar{\mu}_{\bullet,\bullet}$ such that the parameter sets $\{\mu_{rs}\}$ and $\{c, a_r, b_s, \gamma_{rs}\}$ are uniquely determined by each other. Therefore only the μ_{rs} must be estimated.

Due to the normality assumption m.l. clustering is here equivalent to minimizing the total SSQ (1) with respect to $\{\mu_{rs}\}$ and $(\mathscr{A}, \mathscr{B})$. Therefore all statements of Sect. 2 apply and insofar also the clustering criteria (2) and (3) are justified by a probabilistic model (Bock, 1980).

3.3 Maximizing the interaction criterion

Surprisingly it appears that the interaction criterion $G(\mathscr{A}, \mathscr{B}; X)$, (7), can be (approximately) maximized by the same algorithms that have been developed for minimizing the SSQ criterion $Q_{min}(\mathscr{A}, \mathscr{B}; Y)$, (2), if the original data matrix X is suitably transformed before (see also Bock, 1980). In fact:

Theorem 1. *Maximizing the interaction criterion $G(\mathscr{A}, \mathscr{B}; X)$ from (7) is equivalent to minimizing the SSQ clustering criterion $Q_{min}(\mathscr{A}, \mathscr{B}; Y)$ from (2) where the data matrix X has been replaced by the double-centered matrix $Y = (y_{ij})_{I \times J}$ with entries*

$$y_{ij} := x_{ij} - \bar{x}_{i,\bullet} - \bar{x}_{\bullet,j} + \bar{x}_{\bullet,\bullet} \qquad \text{for all } i, j.$$

Proof. It is easily seen that for all r, s:

$$\bar{y}_{A_r \times B_s} = \bar{x}_{A_r \times B_s} - \bar{x}_{A_r,\bullet} - \bar{x}_{\bullet,B_s} + \bar{x}_{\bullet,\bullet} = \hat{\gamma}_{rs}^{(X)}.$$

Therefore the interaction criterion $G(\mathscr{A}, \mathscr{B}; X)$ is identical to the criterion $k(\mathscr{A}, \mathscr{B}; Y)$ from (3). On the other hand, the well-known decomposition formula

$$\sum_{i=1}^{I} \sum_{j=1}^{J} ||y_{ij}||^2 = \underbrace{\sum_{r=1}^{m} \sum_{s=1}^{n} \sum_{\substack{(i,j) \in \\ A_r \times B_s}} ||y_{ij} - \bar{y}_{A_r \times B_s}||^2}_{} + \underbrace{\sum_{r=1}^{m} \sum_{s=1}^{n} |A_r| \cdot |B_s| \cdot ||\bar{y}_{A_r \times B_s}||^2}_{}$$

$$= \quad Q_{min}(\mathscr{A}, \mathscr{B}; Y) \quad + \quad k(\mathscr{A}, \mathscr{B}; Y). \qquad (9)$$

(where the left hand side is constant with respect to $\mathscr{A}, \mathscr{B}$) shows that maximizing the criterion $k(\mathscr{A}, \mathscr{B}; Y)$ is equivalent to minimizing the SSQ criterion $Q_{min}(\mathscr{A}, \mathscr{B}; Y)$ for the double-centered matrix Y. qed

4 Two-way clustering for a contingency table

In this section we consider again a two-way factorial design with two categorical characteristics U and V as in Sect. 3, but here we assume that the entries x_{ij} of the data matrix X are counts N_{ij} and write $X = \mathscr{N} = (N_{ij})_{I \times J}$ in this case. As an example we may consider the N clients (contracts) of a car insurance company, characterized by the profession U of the client and the brand V of the insured car. Then N_{ij} is the number of clients with profession i and car make j. For the company it can make sense to reduce the large numbers of categories I and J to a smaller number m of (profession) classes A_r and a smaller number n of (brand) classes B_s such that profession classes are, on the average, most predictive for the brand class of a client, i.e., with a maximum interaction between both. The resulting classes A_r, B_s and biclusters $A_r \times B_s$ might be the basis for calculating adequate insurance premiums.

In contrast to Sect. 3 where normal distributions were involved, the new scenario is modeled by a random sample of N items (clients) such that N_{ij} is the number of items assigned to the category combination (i, j) (with $\sum_{ij} N_{ij} = N$). Then $\mathscr{N} = (N_{ij})$ has a polynomial distribution $\mathscr{P}ol(N; (p_{ij})_{I \times J})$ with unknown cell probabilities p_{ij} which are typically estimated by $\hat{p}_{ij} := N_{ij}/N$.

In this framework "independence among row and column classes" is modeled by the "hypothesis" H_0:

$$P(A_r \times B_s) = P_U(A_r) \cdot P_V(B_s) \qquad \text{for all } r, s$$

with $P(A_r \times B_s) := \sum_{i \in A_r} \sum_{j \in B_s} p_{ij}$, $P_U(A_r) := \sum_{i \in A_r} \sum_{j=1}^{J} p_{ij}$, $P_V(B_s) := \sum_{i=1}^{I} \sum_{j \in B_s} p_{ij}$, and can be tested, for a fixed bi-partition $(\mathscr{A}, \mathscr{B})$, by the classical χ^2 test. On the other hand, the contrasting idea of "maximum interaction between row and column classes" is interpreted here in the way that the χ^2 test is maximally significant for rejecting H_0, i.e., that the χ^2 test statistics, termed χ^2 *clustering criterion*

$$C(\mathscr{A}, \mathscr{B}) := \sum_{r=1}^{m} \sum_{s=1}^{n} \frac{(\hat{P}(A_r \times B_s) - \hat{P}_U(A_r) \cdot \hat{P}_V(B_s))^2}{\hat{P}_U(A_r) \cdot \hat{P}_V(B_s)} \to \max_{\mathscr{A}, \mathscr{B}} \tag{10}$$

is maximal with respect to the bi-partition $(\mathscr{A}, \mathscr{B})$. Here $\hat{P}$ means the m.l. estimate for the probability distribution P, e.g. with $\hat{P}_{U,V}(A_r \times B_s) = \sum_{i \in A_r} \sum_{j \in B_s} \hat{p}_{ij}$ $= \sum_{i \in A_r} \sum_{j \in B_s} N_{ij}/N$.

In a more general context we note that the χ^2 criterion (10) results as a special case (for $\phi(\lambda) := (\lambda - 1)^2$) from the classical ϕ-divergence measure by Csiszár:

$$C_\phi(\mathscr{A}, \mathscr{B}) := \sum_{r=1}^{m} \sum_{s=1}^{n} \hat{P}_U(A_r) \hat{P}_V(B_s) \cdot \phi\left(\frac{\hat{P}(A_r \times B_s)}{\hat{P}_U(A_r) \hat{P}_V(B_s)}\right) \to \max_{\mathscr{A}, \mathscr{B}} \tag{11}$$

where ϕ is an arbitrary convex function. This *divergence clustering criterion* measures the deviation between the observed probability distribution $\hat{P}$ and the product distribution $\hat{P}_U \cdot \hat{P}_V$ for a given biclustering $(\mathscr{A}, \mathscr{B})$. For $\phi(\lambda) = -\log \lambda$ a Kullback-Leibler clustering criterion results. These criteria have been proposed for clustering by Bock (1983, 1992, 2003, 2004), Celeux et al (1989, χ^2 criterion), Dhillon et al (2003) and Banerjee et al (2005, 2007). Note that the usage of the χ^2 criterion can be justified by theoretical considerations in terms of maximum power, Bahadur efficiency etc. of the χ^2 test (Bock, 1992).

In order to minimize the divergence criterion we may use the classical alternating maximization scheme (*generalized double k-means*): Choose an initial bipartition $\mathscr{A}^{(0)}, \mathscr{B}^{(0)}$ and then alternate between (i) partial maximization with respect to the row partition $\mathscr{A}$ (for fixed $\mathscr{B}$) and (ii) partial maximization with respect to the column partition $\mathscr{B}$ (for fixed $\mathscr{A}$). In order to conduct these partial minimization steps Bock (1992, 2003, 2004) has proposed a k-means-type algorithm that uses class-specific tangents (subgradients) of the convex function ϕ (instead of class means as in the classical SSQ case) and was therefore termed *k-tangent algorithm*. See also Dhillon et al (2003) and Banerjee et al (2005, 2007). For a mixture-type approach see Govaert and Nadif (2010, 2013).

5 Two-way clustering for an object by variable matrix

In the previous sections clustering of rows and columns of the data matrix $X = (x_{ij})_{I \times J}$ was performed in a symmetrical way such that the roles of rows and columns could have been reversed without changing the results. This is different in the case of an object by variable data matrix since, e.g., objects will be independently sampled while variables might be more or less dependent. Also the motivations for grouping objects and variables are different: objects are assembled in groups because they are supposed to behave similarly (with respect to all variables) whereas variables from the same group are supposed to be dependent from each other while independence may hold for variables of different groups. In this last section we sketch two approaches for modeling bi-partition structures for X in the case of I objects and J continuous variables. For more information see, e.g., Vichi (2012); Nadif and Govaert (2010); Govaert and Nadif (2013).

In a probabilistic framework the rows $x_i = (x_{i1}, ..., x_{iJ})'$ of X are considered as a sample of I independent random (column) vectors $X_i = (X_{i1}, ..., X_{iJ})'$ with a distribution that depends on the group A_r of $\mathscr{A} = (A_1, ..., A_m)$ to which object i belongs to. Any clustering $\mathscr{B} = (B_1, ..., B_n)$ of the set of columns $\mathscr{J}$ (with group sizes $b_s := |B_s|$, $s = 1, ..., n$, $\sum_s b_s = J$) is supposed to split the set $\mathscr{J}$ of variables into n mutually independent groups of variables. This also amounts to splitting X_i into n subvectors $X_{i,B_1}, ..., X_{i,B_n}$ such that $X_{i,B_s} \in R^{b_s}$ comprizes the components X_{ij} of X_i that belong to class B_s. For notational convenience we assume here that the ordering of components in X_i is such that all classes $B_1, ..., B_n$ comprize contiguous sets of variables $j \in \mathscr{J}$ such that $X_i = (X'_{i,B_1}, ..., X'_{i,B_n})'$.

A first clustering model is based on the J-dimensional normal distribution:

$$X_i := \begin{pmatrix} X_{i1} \\ \vdots \\ X_{iJ} \end{pmatrix} = \begin{pmatrix} X_{i,B_1} \\ \vdots \\ X_{i,B_n} \end{pmatrix} \sim \mathscr{N}_J(\mu^{(r)}(\mathscr{B}); \Sigma^{(r)}(\mathscr{B})) \quad \text{for } i \in A_r \tag{12}$$

$(r = 1, ..., m)$ where object classes A_r are characterized by class-specific and partitioned expectations $\mu^{(r)}(\mathscr{B}) \in R^J$ and $J \times J$ covariance matrices $\Sigma^{(r)}(\mathscr{B})$ according to

$$\mu^{(r)}(\mathscr{B}) = \begin{pmatrix} \mu_{r,B_1} \\ \vdots \\ \mu_{r,B_n} \end{pmatrix} \qquad \Sigma^{(r)}(\mathscr{B}) = diag(\Sigma_{11}^{(r)}, \cdots \Sigma_{nn}^{(r)}) \tag{13}$$

In particular, we then have, for all $i \in A_r$, that $X_{i,B_s} \sim \mathscr{N}_{b_s}(\mu_{r,B_s}, \Sigma_{ss}^{(r)})$ with independent subvectors X_{i,B_s}, X_{i,B_t} for different column classes B_s and B_t.

While, in principle, m.l. clustering might be possible for this general case, practical applications may concentrate on more parsimonious covariance models, e.g.:

- with independent variables within each group: $\Sigma_{ss}^{(r)} = \sigma_s^{(r)^2} I_{b_s}$ for all s (and then, a fortiori, independence among all J variables);
- with the same variances in all object classes A_r: $\sigma_s^{(r)^2} = \sigma_s^2$ for all r and s;
- with the same variances $\sigma_1^2 = \cdots = \sigma_n^2$ for all groups B_s (then variable groups differ only by the expectation vectors μ_{r,B_s}).

A related mixture model approach is described, e.g., by Nadif and Govaert (2010).

A second modeling approach is based on characteristic subspaces for the variables in B_s, but is only briefly sketched here in a simple case. Let us denote the J column variables of X by $Y_1, ..., Y_J$. We start from the assumption that within each column class B_s, the corresponding random vector Y_{B_s} (that corresponds to the subvector X_{i,B_s} in the matrix X) is generated by a T-dimensional random vector $U^{(s)} := (U_1^{(s)}, ..., U_T^{(s)})'$ such that $Y_{B_s} = \alpha^{(s)} + \sum_{t=1}^{T} \beta_t^{(s)} U_t^{(s)} = \alpha^{(s)} + \beta^{(s)'} U^{(s)}$ is a linear function of the underlying T "factors" or "components" $U_1^{(s)}, ..., U_T^{(s)}$ (which are assumed to be independent, centered and normalized, with $T \le b_s$) with unknown $\alpha^{(s)}$ and coefficients $\beta_t^{(s)}$. Thus, in row i of X, all data subvectors X_{i,B_s} are lying in the same T-dimensional subspace $H^{(s)}$ of R^{b_s} with coordinate vectors $U_{[i]}^{(s)} = (U_{i1}^{(s)}, ..., U_{iT}^{(s)})'$ (typically with $T = 1$

or 2). Typically this subspace will be different for different object groups A_r. Completing the corresponding index r in the previous notation, we obtain the *two-way subspace model*

$$X_{i,B_s} = \alpha_r^{(s)} + \beta_r^{(s)'} U_{[i]}^{(s)} \qquad \text{for } i \in A_r, r = 1,...,m, s = 1,...,n \qquad (14)$$

where the coordinate vectors $U_{[i]}^{(s)}$ are all supposed to be independent. Applying this model (under normal distribution assumptions) to the given data X, we obtain the following *two-way subspace clustering criterion*:

$$R(\mathscr{A},\mathscr{B},\alpha,\beta,u) := \sum_{r=1}^{m} \sum_{i \in A_r} \sum_{s=1}^{n} ||x_{i,B_s} - \alpha_r^{(s)} - \beta_r^{(s)'} u_{[i]}^{(s)}||^2 \to \min_{\mathscr{A},\mathscr{B},\alpha,\beta,u} \qquad (15)$$

which is to be minimized with respect to the parameters and the underlying (factor weighting) vectors $u_{[i]}^{(s)} = (u_{i1}^{(s)},...,u_{iT}^{(s)})' \in R^T$. Essentially this amounts to mn block-specific principal component analyses. After all, the component vectors $u_{[i]}^{(s)}$ can be displayed in R^T and then provide an idea about the configurations of the data within the data blocks $A_r \times B_s$. Similar models and algorithms are surveyed in Vichi (2012); quite generally they provide a remarkable reduction in data complexity in case of a large number J of variables that is reduced here to the dimension nT.

Finally we want to point to the fact that two-way clustering can also be seen in the context of (social) network analysis where we are given, in the simplest case, a data matrix that describes a binary relation among objects (rows) and properties (columns). The problem then consists in constructing blocks of objects (e.g., persons) with a similar behaviour with respect to the properties, and blocks of similarly related properties, all formulated in graphtheoretical terms. Suitable probabilistic and non-probabilistic models and methods are described, e.g., in the seminal publications by Holland and Leinhardt (1981); Anderson et al (1992); Wasserman and Faust (1994); Nowicki and Snijders (2001). Another approach is followed by Harris and Godehardt (1998); Godehardt and Jaworski (2003) and Godehardt et al (2010) who consider, to a given binary relation matrix, the corresponding "intersection graph" for objects and attributes, and analyze its properties in various probabilistic data models.

References

Anderson CJ, Wasserman S, Faust K (1992) Building stochastic blockmodels. Social Networks 14:137–161, DOI 10.1016/0378-8733(92)90017-2

Arabie P, Schleutermann S, Daws J, Hubert L (1988) Marketing applications of sequencing and partitioning of nonsymmetric and/or two-mode matrices. In: Gaul W, Schader M (eds) Data, Expert Knowledge and Decisions, Springer, Berlin, pp 215–224, DOI 10.1007/978-3-642-73489-2_18

Baier D, Gaul W, Schader M (1997) Two-mode overlapping clustering with applications to simultaneous benefit segmentation and market structuring. In: Klar R, Opitz O (eds) Classification and Knowledge Organization, Studies in Classification, Data Analysis, and Knowledge Organization, Springer, Berlin, pp 557–566, DOI 10.1007/978-3-642-59051-1_58

Banerjee A, Merugu S, Dhillon IS, Ghosh J (2005) Clustering with Bregman divergences. The Journal of Machine Learning Research 6:1705–1749

Banerjee A, Dhillon IS, Ghosh J, Merugu S, Modha DS (2007) A generalized maximum entropy approach to Bregman co-clustering and matrix approximation. The Journal of Machine Learning Research 8:1919–1986

Bocci L, Vicari D, Vichi M (2006) A mixture model for the classification of three-way proximity data. Computational Statistics & Data Analysis 50(7):1625–1654, DOI 10.1016/j.csda.2005.02.007

Bock HH (1980) Simultaneous clustering of objects and variables. In: Tomassone R, Amirchahy M, Néel D (eds) Analyse de données et informatique, INRIA, pp 187–203

Bock HH (1983) A clustering algorithm for choosing optimal classes for the chi-squared test. In: Contributed Papers, vol 2, Bull. 44th Session of the International Statistical Institute, pp 758–762

Bock HH (1992) A clustering technique for maximizing ϕ-divergence, noncentrality and discriminating power. In: Schader M (ed) Analyzing and Modeling Data and Knowledge, Studies in Classification, Data Analysis, and Knowledge Organization, Springer, Berlin, pp 19–36, DOI 10.1007/978-3-642-46757-8_3

Bock HH (1996a) Probabilistic methods in cluster analysis. Computational Statistics and Data Analysis 23:5–28

Bock HH (1996b) Probability models and hypothesis testing in partitioning cluster analysis. In: Arabie P, Hubert LJ, De Soete G (eds) Clustering and classification, Studies in Classification, Data Analysis, and Knowledge Organization, World Scientific, Singapore, pp 377–453

Bock HH (2003) Two-way clustering for contingency tables: Maximizing a dependence measure. In: Schader M, Gaul W, Vichi M (eds) Between Data Science and Applied Data Analysis, Studies in Classification, Data Analysis, and Knowledge Organization, Springer, Berlin, pp 143–154, DOI 10.1007/978-3-642-18991-3_17

Bock HH (2004) Convexity-based clustering criteria: Theory, algorithms, and applications in statistics. Statistical Methods and Applications 12(3):293–317, DOI 10.1007/s10260-003-0069-8

Castillo W, Trejos J (2002) Two-mode partitioning: Review of methods and application of tabu search. In: Jajuga K, Sokolowski A, Bock HH (eds) Classification, Clustering, and Data Analysis, Studies in Classification, Data Analysis, and Knowledge Organization, Springer, Berlin, pp 43–51, DOI 10.1007/978-3-642-56181-8_4

Celeux G, Diday E, Govaert G, Lechevallier Y, Ralambondrainy H (1989) Classification automatique des données : environnement statistique et informatique, Dunod, Paris, chap 2.6

Charrad M, Ben Ahmed M (2011) Simultaneous clustering: A survey. In: Kuznetsov SO, Mandal DP, Kundu MK, Pal SK (eds) Pattern Recognition and Machine Intelligence, Lecture Notes in Computer Science, vol 6744, Springer, Berlin, pp 370–375, DOI 10.1007/978-3-642-21786-9_60

Cheng Y, Church GM (2000) Biclustering of expression data. In: Proc. 8th International Conference on Intelligent Systems for Molecular Biology, vol 8, pp 93–103

Cho H, Dhillon IS (2008) Co-clustering of human cancer microarrays using Minimum Sum-Squared Residue co-clustering. IEEE/ACM Transactions on Computational Biology and Bioinformatics 5(3):385–400

Cho H, Dhillon IS, Guan Y, Sra S (2004) Minimum sumsquared residue co-clustering of gene expression data. In: Proc. 4th SIAM International Conference on Data Mining, pp 114–125

Dhillon IS, Mallela S, Modha DS (2003) Information-theoretic co-clustering. In: Proceedings of the Ninth ACM SIGKDD International Conference on Knowledge Discovery and Data Mining, ACM, New York, KDD ’03, pp 89–98, DOI 10.1145/956750.956764

Gaul W, Schader M (1996) A new algorithm for two-mode clustering. In: Bock HH, Polasek W (eds) Data Analysis and Information Systems, Studies in

Classification, Data Analysis, and Knowledge Organization, Springer, Berlin, pp 15–23, DOI 10.1007/978-3-642-80098-6_2

Godehardt E, Jaworski J (2003) Two models of random intersection graphs for classification. In: Schwaiger M, Opitz O (eds) Exploratory Data Analysis in Empirical Research, Studies in Classification, Data Analysis, and Knowledge Organization, Springer, Berlin, pp 67–81, DOI 10.1007/978-3-642-55721-7_8

Godehardt E, Jaworski J, Rybarczyk K (2010) Isolated vertices in random intersection graphs. In: Fink A, Lausen B, Seidel W, Ultsch A (eds) Advances in Data Analysis, Data Handling and Business Intelligence, Studies in Classification, Data Analysis, and Knowledge Organization, Springer, Berlin, pp 135–145, DOI 10.1007/978-3-642-01044-6_12

Govaert G (1995) Simultaneous clustering of rows and columns. Control and Cybernetics 24(4):437–458

Govaert G, Nadif M (2003) Clustering with block mixure models. Pattern Recognition 36:463–473

Govaert G, Nadif M (2005) An EM algorithm for the block mixture model. Pattern Analysis and Machine Intelligence 27(4):643–647, DOI 10.1109/TPAMI.2005.69

Govaert G, Nadif M (2007) Block Bernoulli parsimonious clustering models. In: Brito P, Cucumel G, Bertrand P, de Carvalho F (eds) Selected Contributions in Data Analysis and Classification, Studies in Classification, Data Analysis, and Knowledge Organization, Springer, Berlin, pp 203–212, DOI 10.1007/978-3-540-73560-1_19

Govaert G, Nadif M (2008) Block clustering with Bernoulli mixture models: Comparison of different approaches. Computational Statistics & Data Analysis 52(6):3233–3245, DOI 10.1016/j.csda.2007.09.007

Govaert G, Nadif M (2010) Latent block model for contingency table. Communications in Statistics - Theory and Methods 39(3):416–425, DOI 10.1080/03610920903140197

Govaert G, Nadif M (2013) Co-Clustering. Computing Engineering Series, Wiley, Chichester, UK

Hansohm J (2002) Two-mode clustering with genetic algorithms. In: Gaul W, Ritter G (eds) Classification, Automation, and New Media, Studies in Classification, Data Analysis, and Knowledge Organization, Springer, Berlin, pp 87–93, DOI 10.1007/978-3-642-55991-4_9

Harris B, Godehardt E (1998) Probability models and limit theorems for random interval graphs with applications to cluster analysis. In: Balderjahn I, Mathar

R, Schader M (eds) Classification, Data Analysis, and Data Highways, Studies in Classification, Data Analysis, and Knowledge Organization, Springer, Berlin, pp 54–61, DOI 10.1007/978-3-642-72087-1_6

Holland PW, Leinhardt S (1981) An exponential family of probability distributions for directed graphs. Journal of the American Statistical Association 76(373):33–50, DOI 10.1080/01621459.1981.10477598

Kiers HAL, Vicari D, Vichi M (2005) Simultaneous classification and multidimensional scaling with external information. Psychometrika 70(3):433–460, DOI 10.1007/s11336-002-0998-4

Lazzeroni L, Owen A (2002) Plaid models for gene expression data. Statistica Sinica 12(1):61–86

Li J, Zha H (2006) Two-way Poisson mixture models for simultaneous document classification and word clustering. Computational Statistics & Data Analysis 50(1):163–180, DOI 10.1016/j.csda.2004.07.013

Li T (2005) A general model for clustering binary data. In: Proceedings of the Eleventh ACM SIGKDD International Conference on Knowledge Discovery in Data Mining, ACM, New York, KDD '05, pp 188–197, DOI 10.1145/1081870.1081894

Madeira SC, Oliveira AL (2004) Biclustering algorithms for biological data analysis: A survey. IEEE/ACM Transactions on Computational Biology and Bioinformatics 1(1):24–45, DOI 10.1109/TCBB.2004.2

Martella F, Vichi M (2012) Clustering microarray data using model-based double k-means. Journal of Applied Statistics 39(9):1853–1869, DOI 10.1080/02664763.2012.683172

Martella F, Alfò M, Vichi M (2008) Biclustering of gene expression data by an extension of mixtures of factor analyzers. The International Journal of Biostatistics 4(1)

Martella F, Alfò M, Vichi M (2011) Hierarchical mixture models for biclustering in microarray data. Statistical Modelling 11(6):489–505, DOI 10.1177/1471082X1001100602

Mirkin B, Arabie P, Hubert L (1995) Additive two-mode clustering: The error-variance approach revisited. Journal of Classification 12(2):243–263, DOI 10.1007/BF03040857

Nadif M, Govaert G (2010) Model-based co-clustering for continuous data. In: Draghici S, Khoshgoftaar TM, Palade V, Pedrycz W, Wani MA, Zhu X (eds) Proc. 2010 Ninth International Conference on Machine Learning and Applications (ICMLA'10), IEEE Computer Society, pp 175–180

Nowicki K, Snijders TAB (2001) Estimation and prediction for stochastic block-structures. Journal of the American Statistical Association 96(455):1077–1087, DOI 10.1198/016214501753208735

Rocci R, Vichi M (2008) Two-mode multi-partitioning. Computational Statistics & Data Analysis 52(4):1984–2003, DOI 10.1016/j.csda.2007.06.025

Schepers J, Hofmans J (2009) TwoMP: A MATLAB graphical user interface for two-mode partitioning. Behavior Research Methods 41(2):507–514, DOI 10.3758/BRM.41.2.507

Schepers J, van Mechelen I, Ceulemans E (2006) Three-mode partitioning. Computational Statistics & Data Analysis 51(3):1623–1642, DOI 10.1016/j.csda.2006.06.002

Schepers J, Ceulemans E, Van Mechelen I (2008) Selecting among multi-mode partitioning models of different complexities: A comparison of four model selection criteria. Journal of Classification 25(1):67–85, DOI 10.1007/s00357-008-9005-9

Schepers J, Bock HH, Van Mechelen I (2013) Maximal interaction two-mode clustering. Submitted

Shepard RN, Arabie P (1979) Additive clustering: Representation of similarities as combinations of discrete overlapping properties. Psychological Review 86(2):87–123, DOI 10.1037/0033-295X.86.2.87

Turner HL, Bailey TC, Krzanowski WJ, Hemingway CA (2005) Biclustering models for structured microarray data. IEEE/ACM Trans Computational Biology and Bioinformatics 2(4):316–329

Van Mechelen I, Bock HH, De Boeck P (2004) Two-mode clustering methods: A structured overview. Statistical Methods in Medical Research 13(5):363–394, DOI 10.1191/0962280204sm373ra

Van Rosmalen J, Groenen PJF, Trejos J, Castillo W (2009) Optimization strategies for two-mode partitioning. Journal of Classification 26(2):155–181, DOI 10.1007/s00357-009-9031-2

Vichi M (2001) Double k-means clustering for simultaneous classification of objects and variables. In: Borra S, Rocci R, Vichi M, Schader M (eds) Advances in Classification and Data Analysis, Studies in Classification, Data Analysis, and Knowledge Organization, Springer, Berlin, pp 43–52, DOI 10.1007/978-3-642-59471-7_6

Vichi M (2012) Multimode clustering. In: Paper presented at the Symposium on Learning and Data Science (SLDS 2012), Firenze, Italy

Vichi M, Rocci R, Kiers HA (2007) Simultaneous component and clustering models for three-way data: Within and between approaches. Journal of Classification 24(1):71–98, DOI 10.1007/s00357-007-0006-x

Wasserman S, Faust K (1994) Social network analysis: Methods and applications, vol 8. Cambridge University Press, New York

Wilderjans TF, Depril D, Van Mechelen I (2013) Additive biclustering: A comparison of one new and two existing ALS algorithms. Journal of Classification 30(1):56–74, DOI 10.1007/s00357-013-9120-0

Assessment of Stability in Partitional Clustering Using Resampling Techniques

Hans-Joachim Mucha

Abstract The assessment of stability in cluster analysis is strongly related to the main difficult problem of determining the number of clusters present in the data. The latter is subject of many investigations and papers considering different resampling techniques as practical tools. In this paper, we consider non-parametric resampling from the empirical distribution of a given dataset in order to investigate the stability of results of partitional clustering. In detail, we investigate here only the very popular K-means method. The estimation of the sampling distribution of the adjusted Rand index (ARI) and the averaged Jaccard index seems to be the most general way to do this. In addition, we compare bootstrapping with different subsampling schemes (i.e., with different cardinality of the drawn samples) with respect to their performance in finding the true number of clusters for both synthetic and real data.

1 Introduction

Originally, nonparametric bootstrapping is a statistical method for estimating the sampling distribution of an estimator by sampling with replacement from

Hans-Joachim Mucha
Weierstrass Institute for Applied Analysis and Stochastics (WIAS), 10117 Berlin, Mohrenstraße 39, Germany,
✉ mucha@wias-berlin.de

ARCHIVES OF DATA SCIENCE, SERIES A
KIT SCIENTIFIC PUBLISHING
Vol. 1, No. 1, S. 21–39, 2016

DOI 10.5445/KSP/1000058747/02
ISSN 2363-9881

the original sample (Efron, 1979, 1981). Many authors such as Mammen (1992) derived asymptotic results of parametric bootstrapping. Others presented simulation results of parametric/nonparametric bootstrapping (Efron, 1981; Efron and Tibshirani, 1993).

This very simple technique allows estimation of the sampling distribution of almost any statistic. Bootstrapping falls in the broader class of resampling methods and simulation schemes. Some alternative resampling methods are subsampling (draw a subsample to a smaller size without replacement) and jittering (add noise to every single observation), and a combination of both simulation schemes.

In hierarchical cluster analysis (HCA), we found out that bootstrapping performs best for finding the number of clusters (Mucha and Bartel, 2014, 2015). In all cases (toy and real data), it outperforms subsampling. In subsampling, the choice of the parameter "resampling rate" p causes an additional problem. A subsampling rate of 90% (i.e., $p = 0.9$: this corresponds in some sense to tenfold-cross-validation) or greater performs very bad in HCA methods such as Ward and Average Linkage. The question arises: Is bootstrapping also the best choice for stability investigations of results of partitional clustering?

2 Partitional and hierarchical cluster analysis

A recent survey of partitional and hierarchical clustering algorithms is given by Reddy and Vinzamuri (2014). Here we will emphasize the differences of these two families of cluster analysis methods with respect to the results that have to be assessed by resampling methods. Hierarchical clustering looks fit and proper for resampling because of the (usual) unique and parallel clustering of the I observations into partitions of $K = 2, K = 3, \ldots$ clusters. (Here, a partition $P(I,K)$ is simply the exhaustive partitioning of the set of I observations into K subsets (clusters).) In addition, pairwise distances, the usual starting point of hierarchical cluster analysis, are not affected by bootstrapping/subsampling.

The results of partitional (iterative) clustering methods are dependent on the initial partition into a fixed number of clusters K. That's quite different from hierarchical clustering. In addition, the results of some exchange algorithms are also dependent on the sequence of the observation (Mucha, 2009). For instance, Fig. 1 shows a quite bad result of clustering of a dataset of three two-dimensional randomly generated normal subpopulations. The three Gaussian

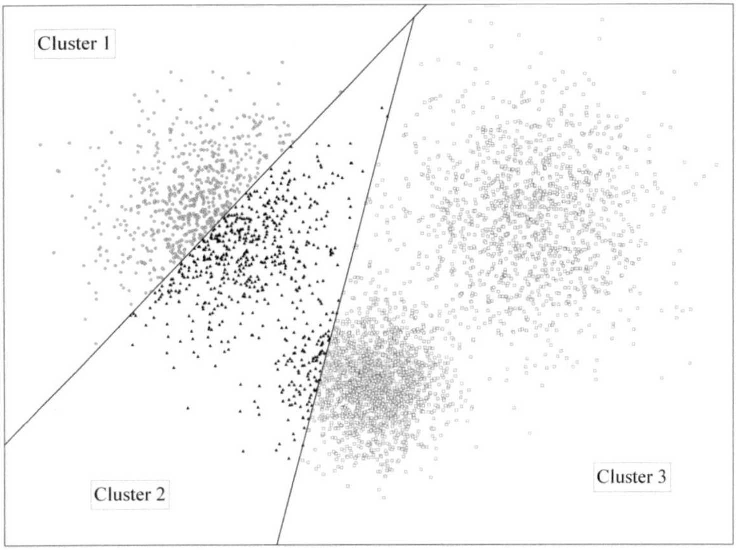

Fig. 1 Result of *Quickcluster* of SPSS applied to 4000 observations.

subpopulations were generated with the following parameters: cardinalities 1100, 1600, and 1300, mean values (-3, 3), (0, 0), and (3, 3), and standard deviations (1, 1), (0.7, 0.7), and (1.2, 1.2). Here the procedure *Quickcluster* of SPSS is applied with the option *running means*: the clusters are updated after each observation is assigned to a new cluster. In this two-dimensional setting, one can check the validity of cluster analysis results visually by eye. In a high-dimensional setting, there is a need for a general validation approach that works in almost all situations (see the next subsection). In this paper, the partitional clustering methods of our software *ClusCorr98* are used (Mucha, 2009). Here, a random access to the observations is realized. This is in order to avoid such bad solutions as shown in Fig. 1. Usually, many different initial partitions, say around 50, are needed to get many different locally optimal solutions. In practice, the best solution is taken for the investigation of stability. Moreover, you have to do this for each K ($K = 2, 3, \ldots$). Finally, you have to do all the things outlined above also for each bootstrap sample (or subsample). Obviously, resampling of partitional clustering looks much more costly in terms of computational complexity than hierarchical clustering. The good news is that some partitional methods such as K-means clustering can work with pairwise distances which are not affected by bootstrapping/subsampling.

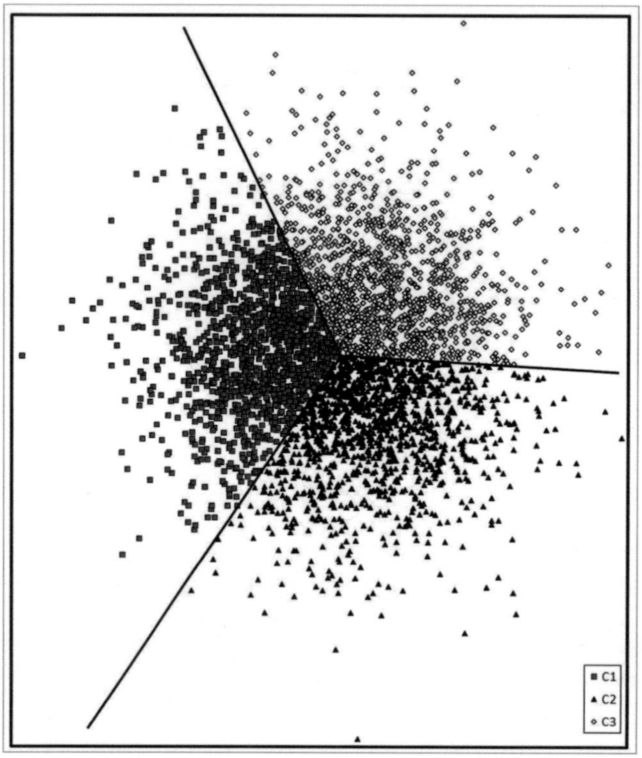

Fig. 2 K-means clustering of the two-dimensional no-structure data into three clusters (marked by black borderlines and color).

Concerning interpretation/comparison of the assessment of stability of two partitions $P(I,K)$ and $P(I,K+1)$ of a hierarchy one has to keep in mind that exactly $K-1$ clusters are identical, i.e., only one cluster is changed when going from $P(I,K)$ to $P(I,K+1)$. That means, theoretically, the lower K the more the stability of the partition $P(I,K)$ depends on the stability of the partition $P(I,K+1)$. This is different from partitional clustering where, usually, all clusters of the two partitions are different.

Even though both clustering techniques, the well-known hierarchical Ward's method and the partitional K-means method, have the same underlying statistical model (Banfield and Raftery, 1993), the results are usually different. Both methods minimize the same criterion (Eq. 1) below but they do this in another way. The K-means clustering method produces the well-known Voronoi tessellation, where the objects have minimum distance to their centroid and,

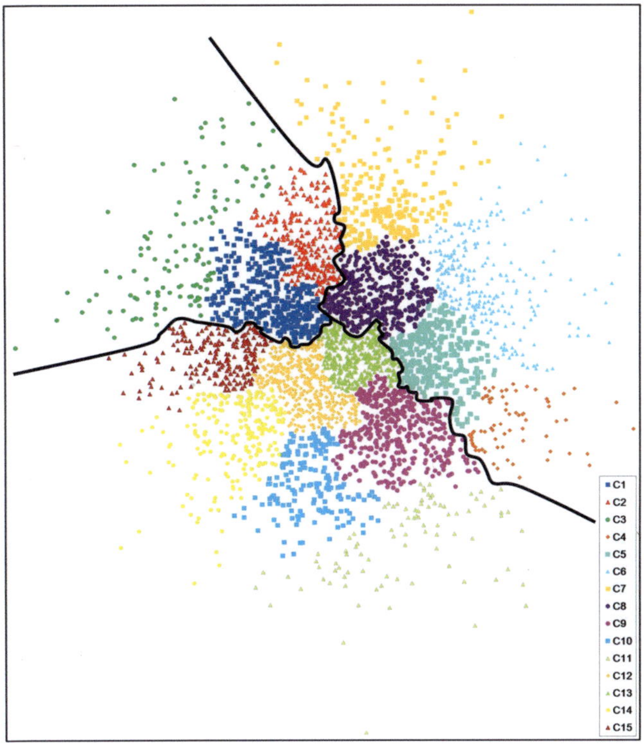

Fig. 3 Ward's clustering of the two-dimensional no-structure data into three clusters (marked by black borderlines) and into 15 clusters (marked by color), respectively.

thus, the borderlines between clusters are hyperplanes as shown in Fig. 2. There are 4000 random generated points in R^2 coming from a standard normally distributed population. In detail: a K-means clustering was done here based on pairwise proximities (squared Euclidean distances, see equations (4) and (5) below). By contrast, the Ward method does not create hyperplanes as borderlines between clusters as illustrated in Fig. 3 for the three cluster solution. Both the hierarchical Ward method and the partitional K-means method minimize the within-cluster sum of squares criterion

$$W_K(\mathbf{G}) = \sum_{k=1}^{K} \mathrm{tr}(\mathbf{W}_k) \tag{1}$$

with respect to a Boolean assignment matrix $\mathbf{G}$ for a fixed K (for details see below).

Herein

$$\mathbf{W}_k = \sum_{i=1}^{I} g_{ik}(\mathbf{x}_i - \overline{\mathbf{x}}_k)(\mathbf{x}_i - \overline{\mathbf{x}}_k)^T \tag{2}$$

is the sample cross-product matrix for the kth cluster $\mathscr{C}_k$ of a given data matrix $\mathbf{X} = (x_{ij})$ consisting of I rows and J columns (variables), and

$$\overline{\mathbf{x}}_k = \frac{1}{g_{.k}} \sum_{i=1}^{I} g_{ik}\mathbf{x}_i \tag{3}$$

is the usual maximum likelihood estimate of expected values in cluster $\mathscr{C}_k$. Further, $g_{.k}$ is the cardinality of cluster $\mathscr{C}_k$, that is, $g_{.k} = \sum_i g_{ik}$.

The Boolean assignment matrix $\mathbf{G}$ formalizes the simplest (elementary) solution to the clustering problem with a fixed number of clusters K: $\mathbf{G} \in \{0,1\}^{I \times K}$ (that is, $\mathbf{G} = (g_{ik})$) with the restriction of uniqueness and exhaustive assignment (completeness) $\sum_{k=1}^{K} g_{ik} = 1$ for every object i. Formally, the mapping is:

$$G : \mathscr{C} \times \{1,2,\ldots,K\} \longrightarrow \{0,1\}$$

with

$$g_{ik} = \begin{cases} 1 \text{ if observation } i \text{ comes from the cluster (subset) } \mathscr{C}_k \\ 0 \text{ otherwise.} \end{cases}$$

Indeed, the cluster mapping G induces a partition $P(I,K) = \{\mathscr{C}_1, \ldots, \mathscr{C}_K\}$ of $\mathscr{C}$. Here, by definition, $\bigcup_{k=1}^{K} \mathscr{C}_k = \mathscr{C}$ and $\mathscr{C}_k \cap \mathscr{C}_l = \emptyset$ for every pair of clusters $\mathscr{C}_k$ and $\mathscr{C}_l$, $k,l = 1,2,\ldots,K, k \neq l$. This cluster mapping yields exactly K clusters (subsets), where the numbering of the clusters is arbitrary because it usually depends on the applied clustering algorithm. Alternatively, let $\mathbf{g} = (g_1, \ldots, g_I)^T$ denote the identifying labels for the clustering and thus for the cluster mapping G, where $g_i = k$ if the ith object $\mathbf{x}_i$ comes from the kth cluster. One can understand $\mathbf{g}$ as a categorical variable or partition variable with K different nominal states $\{1, 2, \ldots, K\}$. Formally, $\mathbf{g} = \mathbf{Ge}$, where the vector $\mathbf{e} = (1,2,3,\ldots,K)^T$ has K entities.

It is well known that the criterion (1) can be written in the following equivalent form without the explicit specification of cluster centers (centroids) $\overline{\mathbf{x}}_k$ (Späth, 1982):

$$W_K(\mathbf{G}) = \sum_{k=1}^{K} \frac{1}{2g_{.k}} \sum_{i=1}^{I} \sum_{h=1}^{I} g_{ik} g_{hk} d_{ih}, \tag{4}$$

and

$$d_{ih} = d(\mathbf{x}_i, \mathbf{x}_h) = (\mathbf{x}_i - \mathbf{x}_h)^T (\mathbf{x}_i - \mathbf{x}_h) = \|\mathbf{x}_i - \mathbf{x}_h\|^2 \tag{5}$$

is the squared Euclidean distance between two observations i and h.

In practice, it is not possible to know how good our best (sub-optimum) result matches both the true (but unknown) classes and the global optimum. We start with many different initial partitions (usually 50), and we select the one that gives the best criterion value. Cluster ensemble methods are another approach in order to find a better cluster analysis result (see, for instance, Minaei-Bidgoli et al, 2014; Fischer and Buhmann, 2003).

3 Resampling techniques in cluster analysis

Nonparametric bootstrapping is resampling taken with replacement from the original data. Equivalently, bootstrapping can be formulated by choosing the following random weights of the observations:

$$m_i = \begin{cases} n \text{ if observation } i \text{ is drawn } n \text{ times} \\ 0 \text{ otherwise.} \end{cases} \tag{6}$$

Here we suppose that the original weights of the observations are $m_i = 1, i = 1, 2, \ldots, I$ ("unit mass"). Then, obviously, $I = \sum_i m_i$ holds in resampling with replacement. Bootstrapping generates multiple observations. When clustering "small" datasets, this can cause problems. The meaning of "small" depends on several factors of influence such as the number of dimensions (variables) and the complexity of the cluster analysis model. Small can be, for instance in the case of simple models such as K-means clustering or Ward's method, a relation $I/K < 5$ with regard to the number of expected clusters K, or a number of observations $I < 20$. In the last situation, soft bootstrapping is recommended by Mucha and Bartel (2014). All statistical methods that make use (directly or indirectly) of weights of the observations can do bootstrapping based on (6). Concerning the K-means method based on pairwise distances, the "centers-free" criterion (4) can be generalized by introducing the weights of the observations to

$$W_K(\mathbf{G}) = \sum_{k=1}^{K} \frac{1}{2M_k} \sum_{i=1}^{I} m_i \sum_{h=1}^{I} g_{ik} g_{hk} m_h d_{ih}. \tag{7}$$

Obviously, it allows a computationally efficient bootstrapping because the pairwise distances (5) remain unchanged in the K-means clustering.

Subsampling is resampling taken without replacement from the original data. It can also be formulated by choosing the following random weights of the observations:

$$m_i^* = \begin{cases} 1 \text{ if observation } i \text{ is drawn randomly} \\ 0 \text{ otherwise.} \end{cases} \quad (8)$$

Here $I > L = \sum_i m_i^*$ holds in resampling without replacement. The parameter $p = L/I$ is needed which causes an additional problem, i.e., setting the cardinality L of the drawn sample. This is different from bootstrapping where no parameter is needed because here the cardinality of the drawn sample always equals I. Below we will investigate subsampling with different p values, say $p = 0.6$ ("Sub60%"), $p = 0.75$ ("Sub75%"), and $p = 0.9$ ("Sub90%"). A practical way out from choosing the parameter p would be to discard multiple points in a bootstrap scheme (named "Boot2Sub" in the investigations below). Concretely, the random bootstrap-weights m_i in (6) have to be modified simply to

$$m_i^* = \begin{cases} 1 \text{ if observation } i \text{ is drawn } n \text{ times} \\ 0 \text{ otherwise.} \end{cases} \quad (9)$$

As a consequence of subsampling via (9), the cardinality of such a subsample "Boot2Sub" is around 63.2% of the I observations (see Efron and Tibshirani, 1997). Clearly, "Boot2Sub" (based on (9)) and bootstrapping (based on (6)) lead to identical results for all the cluster analysis methods that make no use (directly or indirectly) of the weights of the observations m_i such as the hierarchical Single Linkage or Complete Linkage method.

For instance, the resampling method can be used to investigate the variations of the centroids of the clusters, see Mucha and Bartel (2014). As an application, Fig. 5 shows the estimates of the location parameters that are the result of hierarchical Ward's clustering of 250 non-parametric subsamples of the toy dataset presented in Fig. 4. Here three clusters were investigated (for details see Mucha and Bartel, 2014). But, in clustering, the estimation of parameters such as the expected values is not the main task. However, in the case of quantitative data, an estimation of the confidence regions around the cluster centroids can be of interest. The final aim of clustering is the formation of groups either as a partition or a hierarchy of a given set of observations. Therefore, here the focus is on a general investigation of the stability based on partitions. This covers also hierarchies because they can be considered as a set of partitions (Mucha,

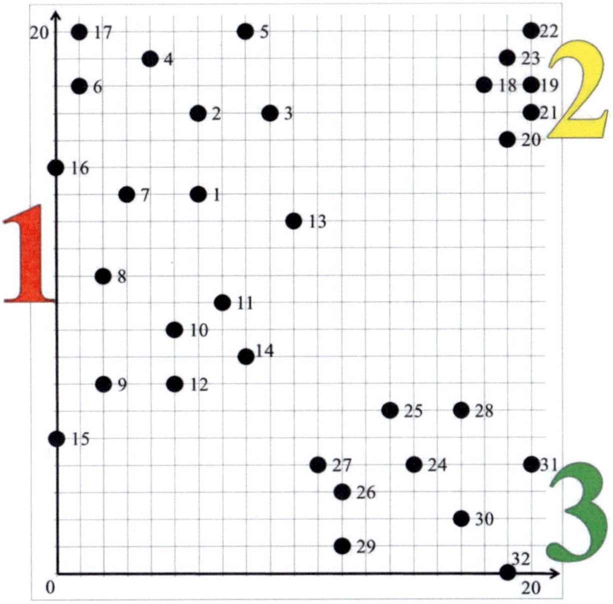

Fig. 4 Plot of the two-dimensional toy dataset divided into three classes by eye. The latter can be found exactly by the partitional K-means clustering. The data values are integers. They can be taken directly from the plot. The observations are numbered.

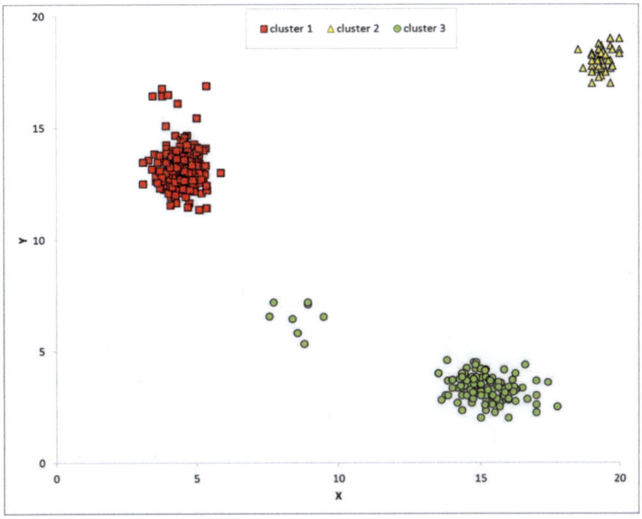

Fig. 5 Plot of the estimates of the location parameter of clusters. They are the result of Ward's HCA of 250 subsamples (75% resampling rate) into three clusters.

2007). To assess the stability of a cluster in the most general way, resampling techniques can be used.

Xiong and Li (2013) investigated many measures of stability with reference to cluster analysis. Here our focus is on two measures, namely the adjusted Rand index (ARI) R and the Jaccard index γ. Why is validation of clustering so important? That is because cluster analysis presents clusters in almost any case. Real clusters should be stable, i.e., they should be confirmed and reproduced to a high degree if the dataset is changed in a non-essential way (Hennig, 2007). Thus, clustering of a randomly drawn sample of a dataset consisting of really well-separated clusters should lead to similar results.

In clustering, usually nothing is known about the true class structure, especially about the number of clusters K. Therefore, the performance or the stability of clustering can not be assessed by counting the rate of misclassifications based on a confusion matrix. However, with the help of non-parametric bootstrapping we are able to operate also on a confusion matrix. It comes from crossing two partitions: the original one and one coming from clustering a "bootstrap" sample. Then the adjusted Rand index or other measures of stability can operate on such an "artificial" confusion matrix. Usually, hundreds of bootstrap samples are needed, see for details (Mucha and Bartel, 2015). Here we work with $B = 250$ bootstrap samples and we take the average (or median) of the B ARI values to come to a final $R_K, K = 2, 3, \ldots$. The maximum R_K gives us an idea about the number of clusters K we are looking for.

In addition to ARI, bootstrapping the Jaccard coefficient can be recommended. The latter assesses the similarity between sets (clusters), for details, see Hennig (2007). It can be used to measure the stability of each individual cluster k by the corresponding Jaccard values γ_k^b with regard to the bootstrap sample $b, b = 1, 2, \ldots, B$. Then we take the average (or median) of the B Jaccard values to come to γ_k that assesses the stability of an individual cluster k. Both the ARI and the averaged Jaccard measure γ_K are recommended for an investigation of the stability of a partition into K clusters. Here, the latter is the average of all Jaccard values γ_k of the K individual clusters of a partition into K clusters. An alternative proposal can be, for instance, a weighted average of all Jaccard values γ_k.

To summarize, bootstrapping of a stability measure is based on an original clustering that is compared many times to corresponding clustering results coming from a bootstrap sample. Concerning more details about bootstrapping a stability index see Mucha (2007); Hennig (2007); Mucha and Bartel (2015).

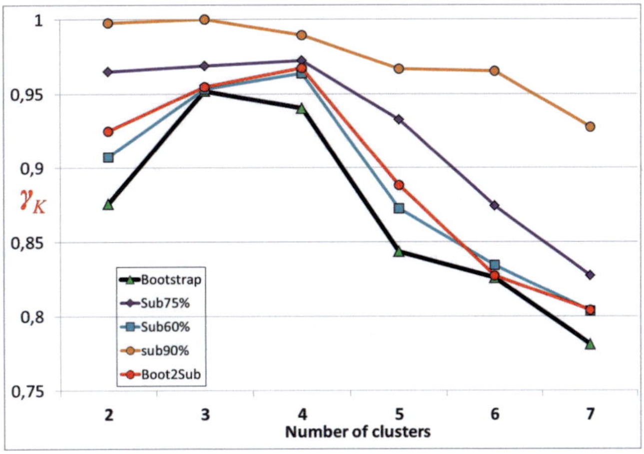

Fig. 6 Jaccard's measures of partitional K-means clustering (shown in Fig. 4) of the toy data.

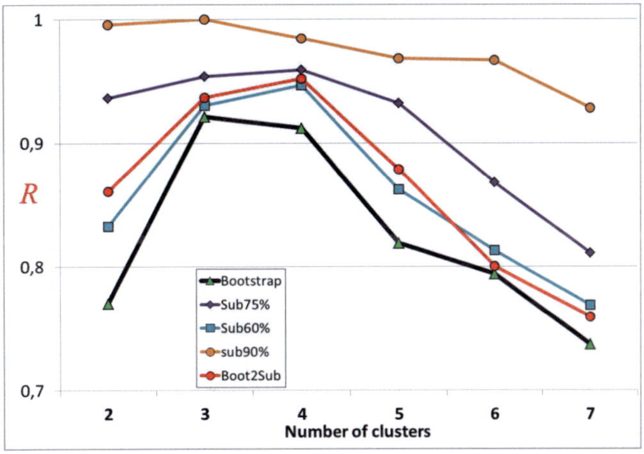

Fig. 7 ARI measures of partitional K-means clustering (shown in Fig. 4) of the toy data.

Other ways of the evaluation of cluster solutions via the bootstrap can be found, for instance, in Fang and Wang (2012), and Dolnicar and Leisch (2010).

4 Bootstrapping versus subsampling in partitional cluster analysis

Fig. 4 introduces a toy dataset consisting of three classes $\mathscr{C}_1 = \{1,2,\ldots,17\}$, $\mathscr{C}_2 = \{18,19,\ldots,23\}$, and $\mathscr{C}_3 = \{24,25,\ldots,32\}$, i.e., it seems to be plausible that there are three classes when looking at the scatterplot. In Fig. 6, different resampling techniques are compared based on the averaged Jaccard measure γ_K for the validation of results of the toy data shown in Fig. 4. The three cluster solution of K-means clustering matches exactly the three classes shown in Fig. 4. Fig. 7 shows similar results as Fig. 6 but based on the ARI R_K.

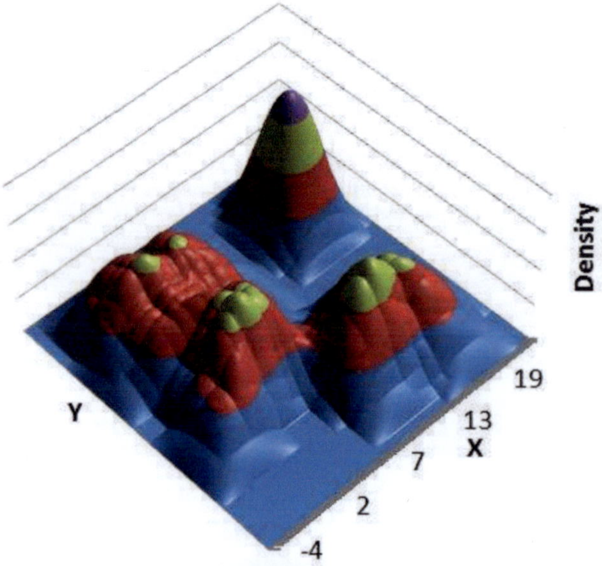

Fig. 8 Plot of the bivariate density estimate of the toy data.

Without much doubt, in this experiment only bootstrapping finds out that there are three clusters. In addition, the ARI "outperforms" Jaccard with respect to the steepest rise when going from $K = 2$ to $K = 3$ clusters. But both present similar results and especially both vote clearly for three clusters and for at most four clusters. The latter because of the steep decrease when going further on to five clusters. Almost all subsampling versions fail in finding the three clusters. In addition, "Sub90%" doesn't indicate any partition clearly. Fig. 8 shows a continuous representation of the toy data. Only class 2 looks homogeneous and well separated (see also Fig. 4), and, maybe, there are more than three peaks.

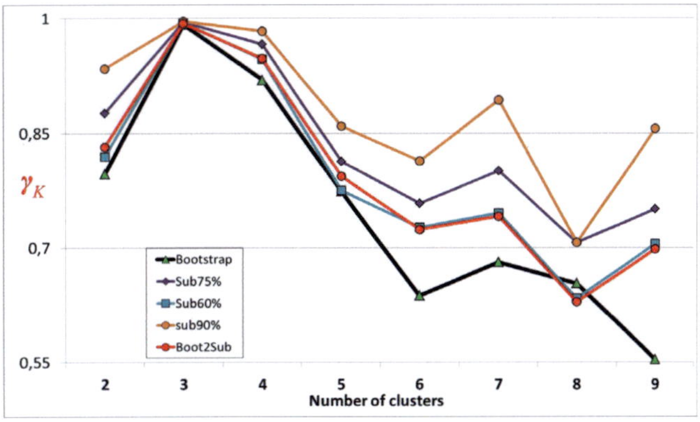

Fig. 9 Jaccard's measures of K-means clustering of the Gaussian data.

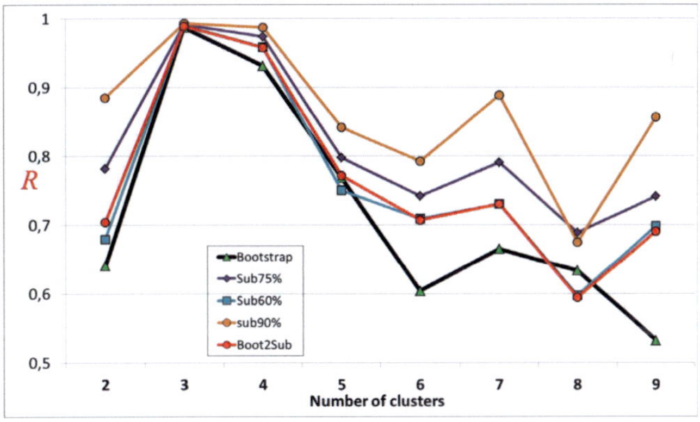

Fig. 10 ARI measures of partitional clustering of the Gaussian data.

Similar to Figs. 6 and 7, Figs. 9 and 10 show the validation results of the partitional K-means clustering of the randomly generated two-dimensional three class data based on the averaged Jaccard measure γ_K and the ARI R_K, respectively. The three Gaussian sub-populations were generated with the following parameters: cardinalities 80, 130, and 90, mean values (-3, 3), (0, 0), and (3, 3), and standard deviations (1, 1), (0.7, 0.7), and (1.2, 1.2). K-means clustering is successful in dividing (decomposing) the data into three subsets: only five errors are counted.

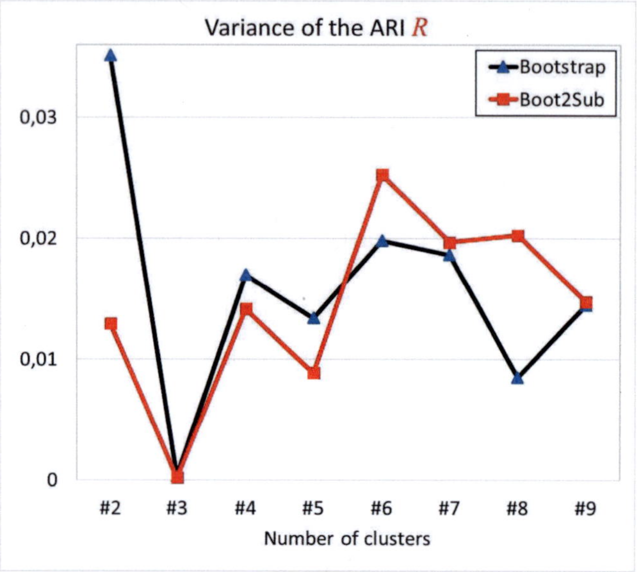

Fig. 11 Variance of ARI measures of partitional clustering of the Gaussian data.

Here bootstrapping performs also best in finding the three classes because it has

1. the maximum value at $K = 3$,
2. the most steeply rising when coming from $K = 2$ and going to $K = 3$, and
3. the most steeply sloping when going further on to $K = 4$.

As before, "Sub90%" performs worst. Subsampling "Boot2Sub" looks most similar to bootstrapping. However, looking at the variances of ARI, say for $K = 2$, bootstrapping has nearly three times more variance (Fig. 11). Bootstrapping looks very instable for the $K = 2$ solution in contrast to "Boot2Sub", and, thus, bootstrapping excludes the wrong solution much clearer.

Next, a real dataset is investigated: the well-known Swiss banknotes data (Flury and Riedwyl, 1988). The data consists of 200 Swiss bank notes based on 6 measurements. There are 100 genuine bank notes and 100 forged ones. Figures 12 and 13 show the validation results of K-means clustering that finds the two classes almost perfectly except for one misclassified observation only. The two true classes are confirmed by both the averaged Jaccard index γ_K and the ARI R_K . The steepest decrease when coming from $K = 3$ and going to $K = 4$ indicates that at most three clusters have a high stability. The latter comes

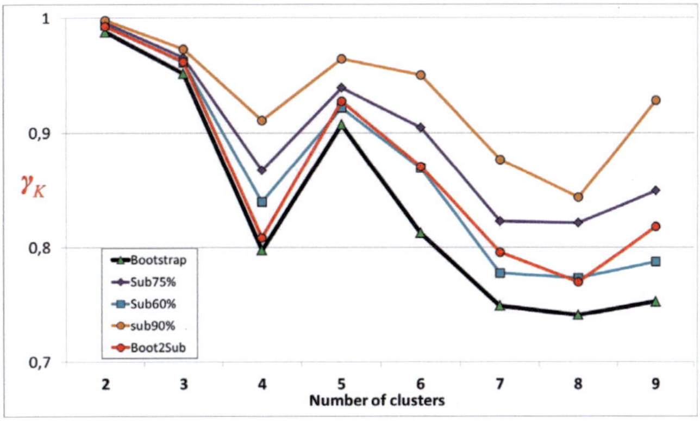

Fig. 12 Averaged Jaccard of the partitional K-means clustering of the Swiss bank notes data.

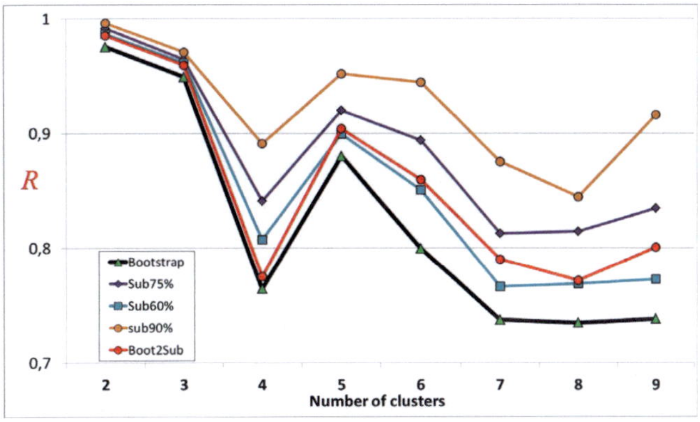

Fig. 13 ARI of the partitional K-means clustering of the Swiss bank notes data.

from the fact that the class of forged bank notes is much more heterogeneous than the class of genuine bank notes (Mucha, 1996). Maybe, the reason for this is that the forged banknotes stem from several different workshops.

The Iris flower data is another well-known real dataset (Fisher, 1936). There are 150 observations that come from three species (classes). One class (species) is easy to find because it looks well separated from the other two in a principle component analysis plot (Mucha, 1992). The other two species are not well separated of each other. 16 errors are counted when using the K-means method with $K = 3$. Fig. 14 and Fig. 15 show the validation results of K-means clustering. The true three classes partition cannot be confirmed by both the averaged Jaccard index and the ARI. The main reason for this failure may be, among others, that K-means clustering is not the appropriate method with an error rate of more than 10%.

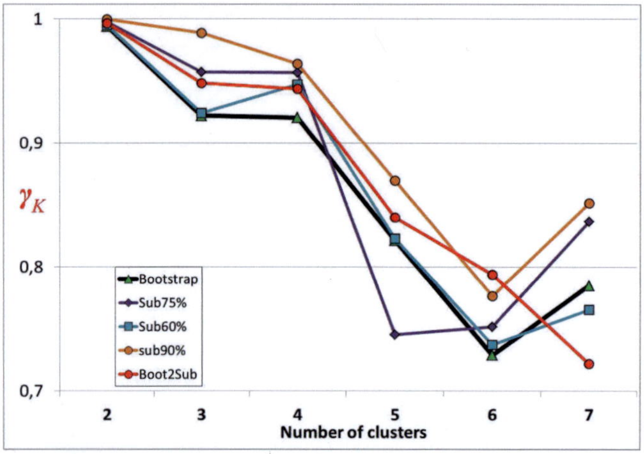

Fig. 14 Averaged Jaccard of the partitional K-means clustering of the Iris data.

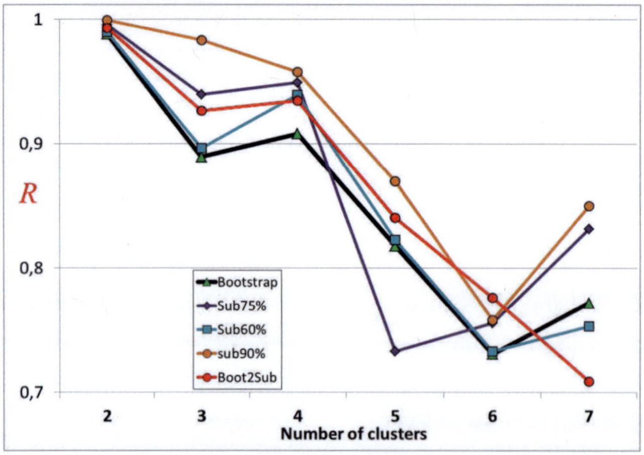

Fig. 15 ARI of the partitional K-means clustering of the Iris data..

5 Summary

In partitional cluster analysis, bootstrapping seems to be also the first choice for both the decision about the number of clusters and the general investigation/assessment of stability. In all cases investigated so far (toy and real data), it outperforms subsampling. It seems to me that multiple observations in bootstrap samples (i.e., observations with mass $m_i > 1$ in (6)) have a great influence for finding the (true) number of clusters. Why? This question has to be answered in the future. The experience of bootstrapping as the winner is similar to the results of hierarchical cluster analysis presented in (Mucha and Bartel, 2014, 2015). But, we investigated here only the very popular K-means method. In subsampling, the choice of the parameter "resampling rate" p causes an additional problem. The simulation results based on a low subsampling rate such as 60% looks similar to bootstrapping. If it necessarily should be subsampling then the recommendation is to take the usual bootstrap scheme but discard multiple observations, i.e., "Boot2Sub". As a consequence, approximately 63.2% of the observations will be presented in such a subsample. The advantage is that no parameter for setting the sample size is necessary.

References

Banfield JD, Raftery AE (1993) Model-based gaussian and non-gaussian clustering. Biometrics 49:803–821

Dolnicar S, Leisch F (2010) Evaluation of structure and reproducibility of cluster solutions using the bootstrap. Marketing Letters 21(1):83–101, DOI 10.1007/s11002-009-9083-4

Efron B (1979) Bootstrap methods: Another look at the jackknife. Annals of Statistics 7(1):1–26, DOI 10.1214/aos/1176344552

Efron B (1981) Nonparametric estimates of standard error: The jackknife, the bootstrap and other methods. Biometrika 68(3):589–599, URL `http://www.jstor.org/stable/2335441`

Efron B, Tibshirani R (1993) An Introduction to the Bootstrap. Chapman and Hall, Boca Raton, USA

Efron B, Tibshirani R (1997) Improvements on cross-validation: The 632+ bootstrap method. Journal of the American Statistical Association 92(438):548–560, DOI 10.1080/01621459.1997.10474007

Fang Y, Wang J (2012) Selection of the number of clusters via the bootstrap method. Computational Statistics and Data Analysis 56(3):468–477, DOI 10.1016/j.csda.2011.09.003

Fischer B, Buhmann JM (2003) Bagging for path-based clustering. IEEE Transactions on Pattern Analysis and Machine Intelligence 25(11):1411–1415, DOI 10.1109/TPAMI.2003.1240115

Fisher RA (1936) The use of multiple measurements in taxonomic problems. Annals of Eugenics 7(2):179–188, URL http://hdl.handle.net/2440/15227

Flury B, Riedwyl H (1988) Multivariate Statistics: A Practical Approach. Chapman and Hall, London

Hennig C (2007) Cluster-wise assessment of cluster stability. Computational Statistics and Data Analysis 52(1):258–271, DOI 10.1016/j.csda.2006.11.025

Mammen E (1992) When Does Bootstrap Work? Asymptotic Results and Simulations. Springer, New York, DOI 10.1007/978-1-4612-2950-6

Minaei-Bidgoli B, Parvin H, Alinejad-Rokny H, Alizadeh H, Punch W (2014) Effects of resampling method and adaptation on clustering ensemble efficacy. Artificial Intelligence Review 41(1):27–48, DOI 10.1007/s10462-011-9295-x

Mucha HJ (1992) Clusteranalyse mit Mikrocomputern. Akademie Verlag, Berlin

Mucha HJ (1996) Cluscorr: Cluster analysis and multivariate graphics under MS EXCEL. In: Mucha HJ, Bock HH (eds) Classification and Clustering: Models, Software and Applications, 10, WIAS, Berlin, pp 97–106

Mucha HJ (2007) On validation of hierarchical clustering. In: Decker R, Lenz HJ (eds) Advances in Data Analysis, Studies in Classification, Data Analysis, and Knowledge Organization, Springer, Berlin, pp 115–122, DOI 10.1007/978-3-540-70981-7_14

Mucha HJ (2009) Cluscorr98 for Excel 2007: Clustering, multivariate visualization, and validation. In: Mucha HJ, Ritter G (eds) Classification and Clustering: Models, Software and Applications, 26, WIAS, Berlin, pp 14–41

Mucha HJ, Bartel HG (2014) Soft bootstrapping in cluster analysis and its comparison with other resampling methods. In: Spiliopoulou M, Schmidt-Thieme L, Janning R (eds) Data Analysis, Machine Learning and Knowledge Discovery, Springer, Berlin, pp 97–104, DOI 10.1007/978-3-319-01595-8_11

Mucha HJ, Bartel HG (2015) Resampling techniques in cluster analysis: Is subsampling better than bootstrapping? In: S Krolak-Schwerdt BL, Böhmer M (eds) European Conference on Data Analysis, Springer, Berlin

Reddy CK, Vinzamuri B (2014) A survey of partitional and hierarchical clustering algorithms. In: Aggarwal CC, Reddy CK (eds) Data Clustering. Algorithms and Appications, Chapman and Hall/CRC, Boca Raton, USA, pp 87–126

Späth H (1982) Cluster analysis algorithms for data reduction and classification of objects. Computers and their applications, Ellis Horwood, Chichester

Xiong H, Li Z (2013) Clustering validation measures. In: Aggarwal CC, Reddy CK (eds) Data Clustering. Algorithms and Applications, Chapman and Hall/CRC, Boca Raton, USA, pp 571–605

Learning Conditional Lexicographic Preference Trees

Michael Bräuning and Eyke Hüllermeier

Abstract We introduce a generalization of lexicographic orders and argue that this generalization constitutes an interesting model class for preference learning in general and ranking in particular. We propose a learning algorithm for inducing a so-called conditional lexicographic preference tree from a given set of training data in the form of pairwise comparisons between objects. Experimentally, we validate our algorithm in the setting of multipartite ranking.

1 Introduction

Preference learning is an emerging subfield of machine learning that has received increasing attention in recent years (Fürnkranz and Hüllermeier, 2011). A specific though important special case of preference learning is "learning to rank", that is, the learning of models that can be used to predict preferences in the form of rankings of a set of alternatives (Cohen et al, 1999; Dekel et al, 2003). Ranking problems are often reduced to problems of a simpler type, such as learning a value function that assigns scores to alternatives (with better

Michael Bräuning
Philipps-University Marburg,
✉ braeunim@mathematik.uni-marburg.de

Eyke Hüllermeier
University of Paderborn,
✉ eyke@upb.de

ARCHIVES OF DATA SCIENCE, SERIES A
KIT SCIENTIFIC PUBLISHING
Vol. 1, No. 1, S. 41–55, 2016

DOI 10.5445/KSP/1000058747/03
ISSN 2363-9881

alternatives having higher scores) or learning a binary predicate that compares pairs of alternatives (Hüllermeier et al, 2008). While the former approach is close to regression, the latter is in the realm of classification learning.

Another approach to learning ranking functions is to proceed from specific model assumptions, that is, assumptions about the structure of the sought preference relations. This approach is less generic than the previous one, as it strongly depends on the concrete assumptions made. On the other hand, it typically offers the advantage of being more easily understandable and interpretable. As an example, let us mention CP-networks, that is, the representation of conditional dependence and independence of preference statements under a ceteris paribus (all else being equal) interpretation (Boutilier et al, 2004). Those preferences are encoded as a graph, in which each node is annotated with a preference table. Another example is lexicographic orders that are widely accepted as a plausible representation of (human) preferences (Schmitt and Martignon, 2006), especially in complex decision making domains (Ahlert, 2008). Here, the assumption is that the target ranking of a set of alternatives, each one described in terms of multiple attributes, can be represented as a lexicographic order.

From a machine learning point of view, assumptions of the above type can be seen as an *inductive bias* restricting the hypothesis space. Provided the bias is correct, this is clearly an advantage, as it may simplify the learning problem. On the other hand, an overly strong bias may prevent the learner from approximating the target ranking sufficiently well. For example, while being plausible in some situations, the assumption of a lexicographic order will be too restrictive for many applications.

In this paper, we therefore present a method for learning generalized lexicographic orders. While still being simple and easy to understand, the model class we consider relaxes some of the assumptions of a proper lexicographic order. More specifically, we increase flexibility thanks to two extensions of conventional lexicographic orders:

- First, we allow for *conditioning* (Booth et al, 2009, 2010): The importance of attributes as well as the preferences for the values of an attribute may depend on the values of other variables preceding that one in the underlying variable order.
- Second, we allow for *grouping* (Wilson, 2009): Several (one-dimensional) variables can be grouped into a single high-dimensional variable, and preferences can be specified on the Cartesian product of the corresponding domains.

The remainder of this paper is organized as follows. In the next section, we give a brief overview of related work. In Sect. 3, we introduce generalized lexicographic orders and the notion of conditional lexicographic preference trees. In Sect. 4, we present an algorithm for learning such preference models from data. An experimental study is presented in Sect. 5, prior to concluding the paper in Sect. 6.

2 Related Work

The use of lexicographic orders in preference modeling has already been considered in the seventies of the last century (Fishburn, 1974), whereas in machine learning, this type of structure has attracted attention only recently. Flach and Matsubara developed a lexicographic ranker called LexRank , using a linear preference ordering on attributes derived by the odds ratio (Flach and Matsubara, 2007, 2008). Experimentally, they show that LexRank is competitive to decision trees and naive Bayes in terms of ranking performance.

Further work on learning lexicographic orders was done by Schmitt and Martignon (2006), Dombi et al (2007), and Yaman et al (2008). However, these works are based on rather simplistic assumptions. More general models were studied by Booth et al (2009, 2010), and in fact, important parts of our approach (such as conditional importance of attributes and conditional preferences on attribute values) are inspired by these models. Their work remains rather theoretical, however, without a practical realization in terms of an implementation of algorithms or an experimental study with real data.

3 Generalized Lexicographic Orders

Formally, we proceed from an attribute-value representation of decision alternatives or objects, i.e., an object is represented as a vector

$$o \in \mathscr{O} = \mathscr{D}(V) = \mathscr{D}(A_1) \times \ldots \times \mathscr{D}(A_n),$$

where $V = \{A_1, \ldots, A_n\}$ is the set of attributes (variables) and $\mathscr{D}(A_i)$ is the domain of attribute A_i. For a subset $A = \{A_{i_1}, \ldots, A_{i_k}\} \subset V$ of attributes we define $\mathscr{D}(A) = \mathscr{D}(A_{i_1}) \times \ldots \times \mathscr{D}(A_{i_k})$.

An *assignment* or *instantiation* of a subset $A \subseteq V$ of attributes is an element $a \in \mathscr{D}(A)$; an assignment is called *complete* if $A = V$, otherwise it is called *partial*. For an object $o \in \mathscr{O}$ and a subset $A \subset V$, we denote by $o[A]$ the projection of o from $\mathscr{D}(V)$ to $\mathscr{D}(A)$; if $A = \{A_k\}$ is a single attribute, we also write $o[k]$ instead of $o[\{A_k\}]$.

A lexicographic order on $\mathscr{O}$ is a total order $\succ$ defined in terms of

- a total order $\sqsupset$ on V, i.e., a ranking of the attributes,
- a total order $\sqsupset_i$ on each attribute domain $\mathscr{D}(A_i)$.

More specifically, $o^* \succ o$ (suggesting that o^* is preferred to o) if and only if there exists a $k \in \{1,\dots,n\}$ such that

$$\left(o^*[k] \sqsupset_k o[k]\right) \wedge \Big(\left(A_i \sqsupset A_k\right) \Rightarrow \left(o^*[i] = o[i]\right)\Big)$$

for all $i \in \{1,\dots,n\}$. The relations $\sqsupset_i$ indicate preference on individual attributes: $a \sqsupset_i b$ means that, for $a, b \in \mathscr{D}(A_k)$, a is preferred to b as a value for attribute A_i. Moreover, the relation $\sqsupset$ reflects the importance of attributes: $A_i \sqsupset A_j$ means that attribute A_i is more important than A_j, whence the former is considered prior to the latter. Without loss of generality, we shall subsequently assume that $A_1 \sqsupset A_2 \sqsupset \cdots \sqsupset A_n$ (unless otherwise stated).

3.1 Conditional preferences on attribute values

Conventional lexicographic orders assume that preferences $\sqsupset_k$ on attribute domains are independent of each other. Needless to say, this assumption is often violated in practice. For example, although it is possible that a person prefers red wine to white wine *in general*, it is also plausible that her preference for wine may depend on the main dish: red is preferred to white in the case of meat, whereas white is preferred to red in the case of fish.

In order to capture attribute dependencies of that type, the preference relations $\sqsupset_k$ can be conditioned on the values of the attributes A_j preceding A_k in the order $\sqsupset$ (Booth et al, 2009, 2010). That is, $\sqsupset_k$ is now replaced by a set of strict orders

$$\left\{ \sqsupset_k^{(a_1,\dots,a_{k-1})} \,|\, (a_1,\dots,a_{k-1}) \in \mathscr{D}(\{A_1,\dots,A_{k-1}\}) \right\}$$

Moreover, the order relation $\succ$ on $\mathscr{O}$ is then defined as follows: $o^* \succ o$ for $o^* = (a_1^*,\dots,a_n^*)$ and $o = (a_1,\dots,a_n)$ if and only if there exists a $k \in \{1,\dots,n\}$ such that

$$\Big(\forall i \in \{1,\ldots,k-1\} : a_i^* = a_i\Big) \wedge \big(a_k^* \sqsupset_k^{(a_1,\ldots,a_{k-1})} a_k\big).$$

3.2 Conditional attribute importance

Going one step further, one may assume that the values of the first attributes in the attribute order $\sqsupset$ do not only influence the preferences on the values of the attributes that follow, but also the importance of the attributes themselves (Booth et al, 2009, 2010). Thus, we are no longer dealing with a lexicographic order in the sense that $\sqsupset$ defines a *sequence* of the attributes V according to their importance. Instead, we are dealing with a *tree-like* structure. This structure is defined by the following (choice) function:

$$A = C\Big((A_{i_1}, A_{i_2}, \ldots, A_{i_k}), (a_{i_1}, a_{i_2}, \ldots, a_{i_k})\Big),$$

where $(A_{i_1}, A_{i_2}, \ldots, A_{i_k}) \in V^k$ is a sequence of attributes (such that $A_{i_j} \neq A_{i_k}$ for $j \neq k$) and $a_{i_j} \in \mathscr{D}(A_{i_j})$ for all $j \in \{1,\ldots,k\}$. Moreover, $A \in V \setminus \{A_{i_1},\ldots,A_{i_k}\}$ is the most important attribute given that $A_{i_j} = a_{i_j}$ for all $j \in \{1,\ldots,k\}$.

3.3 Variable grouping

Another extension consists of grouping several variables, that is, to allow the expression of preferences on *attribute tuples* instead of single attributes only (Wilson, 2009). Formally, this means selecting an index set $\mathscr{I} \subseteq \{1,\ldots,n\}$ and defining a total order relation $\sqsupset_{\mathscr{I}}$ on the Cartesian product $\mathscr{D}(V_{\mathscr{I}})$ of the domains $\mathscr{D}(A_i)$, $i \in \mathscr{I}$.

Note that the possibility of variable grouping significantly increases the expressivity of the model class. In particular, by taking $\mathscr{I} = \{1,\ldots,n\}$, it is possible to define every order on $\mathscr{D}(V)$, that is, to sort the set of alternatives in any way. Since this level of expressivity is normally not desirable, it is reasonable to restrict to variable grouping of order g_{max}, meaning to impose the constraint $|\mathscr{I}| \leq g_{max}$ for a fixed $g_{max} \leq n$.

3.4 *Conditional lexicographic preference trees*

Combining the generalizations discussed above, we end up with what we call a Conditional Lexicographic Preference Tree (CLPT). Graphically, this is a tree structure in which

- every node is labeled with a subset of attributes $V_{\mathscr{I}}$ and a total order on the Cartesian product $\mathscr{D}(V_{\mathscr{I}})$ of the corresponding attribute domains $\mathscr{D}(A_i)$, $i \in \mathscr{I}$;
- there is one outgoing edge (descendant node) for each value $o[V_{\mathscr{I}}] \in \mathscr{D}(V_{\mathscr{I}})$;
- every attribute $A_i \in V$ occurs at most once on each branch from the root of the tree to a leaf node (i.e., the index sets $\mathscr{I}$ along a branch are disjoint).

We call a CLPT *complete* if every attribute $A_i \in V$ occurs exactly once on each branch from the root of the tree to a leaf node (i.e., the index sets $\mathscr{I}$ along a branch form a partition of $\{1,\ldots,n\}$).

A (complete) CLPT can be thought of as defining an order relation on $\mathscr{O}$ through recursive refinement of a weak order $\succeq$, that is, by refining an order relation with tie groups in a recursive manner (in the following, $\sim$ and $\succ$ denote, respectively, the symmetric and asymmetric part of $\succeq$):

- One starts with a single equivalence class (tie group), i.e., $o^* \sim o$ for all $o^*, o \in \mathscr{O}$.
- Let the root of the CLPT be labeled with the attribute set $V_{\mathscr{I}}$, and let $\sqsupset_{\mathscr{I}}$ denote the corresponding order on $\mathscr{D}(V_{\mathscr{I}})$. The current order $\succeq$ is then refined by letting $o^* \succ o$ whenever $o^*[V_{\mathscr{I}}] \sqsupset_{\mathscr{I}} o[V_{\mathscr{I}}]$; otherwise, if $o^*[V_{\mathscr{I}}] = o[V_{\mathscr{I}}]$, then o^* and o remain tied.
- Thus, a linear order of tie groups (equivalence classes) is produced.
- Each equivalence class (represented by a value $a \in \mathscr{D}(V_{\mathscr{I}})$) is then recursively refined by the subtree the objects of this equivalence class are passed to.

Note that, if the CLPT is complete, the order relation $\succeq$ eventually produced is a total order $\succ$.

4 Learning CLPTs

In this section, we outline a method for inducing a CLPT from training data

$$\mathscr{T} = \{(o_i^*, o_i)\}_{i=1}^{N} \tag{1}$$

that consists of a set of object pairs $(o_i^*, o_i) \in \mathscr{O}^2$, suggesting that o_i^* is preferred to o_i. Roughly speaking, this means finding a CLPT whose induced order relation $\succeq$ on $\mathscr{O}$ is as much as possible in agreement with the pairwise preferences in $\mathscr{T}$ (without overfitting the training data). The induced order relation $\succeq$ is a total order $\succ$ if the CLPT is complete.

4.1 Performance and evaluation measures

In order to evaluate the predictive performance of a CLPT, there is a need to compare the order relation $\succeq$ (with asymmetric part $\succ$) induced by this model with a ground truth order $\succ^*$. As will be seen below, the same measures can be used to fit a CLPT to a given set of training data (1) during the training phase. In this case, the "ground truth" is not a total order but a set of pairwise comparisons between objects. Since a total order $\succ^*$ can be decomposed into (a quadratic number of) such comparisons, too, we can assume (without loss of generality) that we compare $\succeq$ with a set $\mathscr{T}$ of pairs $(o^*, o) \in \mathscr{O}^2$, suggesting that o^* should be ranked higher than o.

Inspired by the corresponding notions introduced in Cheng et al (2010), we define two performance measures of *correctness* and *completeness*, respectively, as follows:

$$\mathrm{CR}(\succeq, \mathscr{T}) = \frac{|C| - |D|}{|C| + |D|}, \tag{2}$$

$$\mathrm{CP}(\succeq, \mathscr{T}) = \frac{|C| + |D|}{|\mathscr{T}|}, \tag{3}$$

where

$$C = \left\{(o^*, o) \in \mathscr{T} \,|\, o^* \succ o\right\},$$
$$D = \left\{(o^*, o) \in \mathscr{T} \,|\, o \succ o^*\right\}.$$

Note that $\mathrm{CR}(\succeq, \mathscr{T})$ assumes values between -1 (complete disagreement) and $+1$ (complete agreement), while $\mathrm{CP}(\succeq, \mathscr{T})$ ranges between 0 (no comparisons) and 1 (full comparison).

4.2 A greedy learning procedure

We implement an algorithm for learning a CLPT as a (greedy) search in the space of tree structures based on the greedy algorithms presented by Schmitt and Martignon (2006) as well as Booth et al (2009, 2010). This is done by constructing the tree from the root to the leaves in a recursive manner. In each step of the recursion, a new node is created with an associated subset $V_{\mathscr{I}}$ of attributes, where $|V_{\mathscr{I}}| \leq g_{max}$, and a total order $\sqsupset_{\mathscr{I}}$ on $\mathscr{D}(V_{\mathscr{I}})$.

4.2.1 Creating a node

The problem to be solved in each recursion is the following: Given a set of pairwise comparisons $\mathscr{T}$ and a set $V' \subseteq V$ of attributes still available, select the most suitable subset $V_{\mathscr{I}} \subseteq V'$ and an order $\sqsupset_{\mathscr{I}}$. Following a greedy strategy, we choose $(V_{\mathscr{I}}, \sqsupset_{\mathscr{I}})$ so as to maximize correctness (2), using completeness (3) as a second criterion to break ties. In the (unlikely) event of both correctness and completeness having ties, the first subset $V_{\mathscr{I}}$ and order $\sqsupset_{\mathscr{I}}$ identified are selected.

The selection of an attribute subset $V_{\mathscr{I}}$ can be done through exhaustive search if its size is sufficiently limited, i.e., if the upper bound g_{max} is small. Otherwise, a complete enumeration of all possibilities may become too expensive. Moreover, for each candidate subset $V_{\mathscr{I}}$, a total order $\sqsupset_{\mathscr{I}}$ needs to be determined. Again, all such orders can be tried if $\mathscr{D}(V_{\mathscr{I}})$ is not too large. Otherwise, heuristic ranking procedures such as a Borda count can be used (counting the number of "wins" and "losses" of each value $a \in \mathscr{D}(V_{\mathscr{I}})$ in the training data $\mathscr{T}$ and sorting according to the difference).

4.2.2 Limiting the number of candidate subsets

In order to avoid a complete enumeration of all candidate subsets $V_{\mathscr{I}}$ of size $\leq g_{max}$, we combine a greedy search with a kind of lookahead procedure: We provisionally create a node by selecting a single attribute instead of a subset, i.e., we tentatively set g_{max} to 1; apart from that, exactly the same selection procedure (as outlined above) is applied. This step is repeated g_{max} times, thereby producing a subtree of depth g_{max}. Let $V^* \subseteq V$ denote the subset of attributes that occur in this subtree, i.e., that are chosen in at least one of the nodes. Then, as candidate subsets $V_{\mathscr{I}}$, we only try subsets V^*, i.e., subsets $V_{\mathscr{I}} \subseteq V^*$ such that $|V_{\mathscr{I}}| \leq g_{max}$. Obviously, the underlying assumption is that an attribute that has not been chosen in any of the g_{max} steps is not important at this point.

4.2.3 Recursion

Once an optimal subset $V_{\mathscr{I}}$ has been chosen, the training examples (o^*, o) with $o^*[V_{\mathscr{I}}] \neq o[V_{\mathscr{I}}]$ are removed from $\mathscr{T}$ (since they are sorted at this node). Moreover, for each value $a \in \mathscr{D}(V_{\mathscr{I}})$, a data set

$$\mathscr{T}_a = \left\{ (o^*, o) \in \mathscr{T} \,|\, o^*[V_{\mathscr{I}}] = o[V_{\mathscr{I}}] = a \right\}$$

is created and passed to the corresponding successor node (together with $V' \setminus V_{\mathscr{I}}$ as the attributes that have not been used so far). The same recursive procedure is then applied to each of these successor nodes.

4.2.4 Initialization and termination

The learning procedure is called with the original training set $\mathscr{T}$ and the full set V of attributes as candidates. The recursion terminates if no attribute is left ($V' = \emptyset$) or if the set of training examples is empty ($\mathscr{T} = \emptyset$). A description of the basic algorithm in the form of pseudocode is provided in Algorithm 1.[1]

4.2.5 CLeRa

We call the algorithm outlined above *CLeRa*, which is short for Conditional Lexicographic Ranker . The CLPT induced by CLeRa can be used to compare new object pairs $\{o^*, o\} \subset \mathscr{O}$. To this end, the tuple is submitted to the root and propagated through the tree until either a leaf node is reached or a node at which $o^*[V_{\mathscr{I}}] \neq o[V_{\mathscr{I}}]$; in this case, $o^* \succ o$ is decided if $o^* \sqsupset_{\mathscr{I}} o$ and $o \succ o^*$ if $o \sqsupset_{\mathscr{I}} o^*$. Otherwise, if $o^*[V_{\mathscr{I}}] = o[V_{\mathscr{I}}]$ in all nodes traversed by the two objects, then $o^* \sim o$.

Given not only a pair but a complete set of objects to be ranked, the pairwise comparison realized by the CLPT can be embedded in any standard sorting

[1] The pseudocode does not consider the lookahead procedure.

Algorithm 1: CLeRa

Input : training data $\mathscr{T}$, set of attributes V, maximal grouping size g_{max}
Output : CLPT ct

$ct \leftarrow \emptyset, V' \leftarrow V, \mathscr{I}' \leftarrow \{1,\ldots,n\}$
if $\mathscr{T} \neq \emptyset$ && $V' \neq \emptyset$ **then**
  $I' \leftarrow \emptyset$, $CR \leftarrow 0$, $CP \leftarrow 0$
  for $\mathscr{I} \subseteq \mathscr{I}'$, $|\mathscr{I}| \leq g_{max}$ **do**
    determine $\sqsupset_{\mathscr{I}}$ on $\mathscr{D}(V_{\mathscr{I}})$ maximally consistent with $\mathscr{T}$
    compute $\mathrm{CR}(\sqsupset_{\mathscr{I}}, \mathscr{T})$ and $\mathrm{CP}(\sqsupset_{\mathscr{I}}, \mathscr{T})$
    if $\mathrm{CR}(\sqsupset_{\mathscr{I}}, \mathscr{T}) = \mathrm{CR}$ && $\mathrm{CP} < \mathrm{CP}(\sqsupset_{\mathscr{I}}, \mathscr{T})$ **then**
      $\mathrm{CP} \leftarrow \mathrm{CP}(\sqsupset_{\mathscr{I}}, \mathscr{T})$
      $I' \leftarrow \mathscr{I}$
    else if $\mathrm{CR}(\sqsupset_{\mathscr{I}}, \mathscr{T}) > \mathrm{CR}$ **then**
      $\mathrm{CR} \leftarrow \mathrm{CR}(\sqsupset_{\mathscr{I}}, \mathscr{T})$
      $\mathrm{CP} \leftarrow \mathrm{CP}(\sqsupset_{\mathscr{I}}, \mathscr{T})$
      $I' \leftarrow \mathscr{I}$
  $\mathscr{I}' \leftarrow \mathscr{I}' \backslash I'$
  $V' \leftarrow V' \backslash V_{I'}$
  remove every $(o, o') \in \mathscr{T}$ decided by $\sqsupset_{I'}$
  add node $(V_{I'}, \sqsupset_{I'})$ to ct
  for $a \in \mathscr{D}(V_{\mathscr{I}})$ **do**
    $\mathscr{T}_a = \left\{(o^*, o) \in \mathscr{T} \,|\, o^*[V_{\mathscr{I}}] = o[V_{\mathscr{I}}] = a\right\}$
    return CLeRa$[\mathscr{T}_a, V', g_{max}]$
return ct

algorithm, such as insertion sort. Note that, since $o^* \sim o$ is possible in a pairwise comparison, the result of the sorting procedure will in general only be a weak order $\succeq$.

5 Experimental Results

We evaluate our approach on 15 benchmark data sets from the Statlog and the UCI repository (Asuncion and Newman, 2007). These data sets, which define binary or ordinal classification problems, were pre-processed as follows: numerical attributes and attributes with more than five values were discretized into four values using equal frequency binning. Moreover, instances with missing values were neglected.

The learning problem we consider is multipartite ranking (Fürnkranz et al, 2009): Given a set of test instances $X \subset \mathscr{O}$, the goal is to predict a ranking $\succeq$ that agrees with the (ordered) class labels of these instances. Formally, this agreement is measured in terms of the so-called C-index, which can be seen as an extension of the area under the ROC curve (AUC):

$$C = \frac{1}{\sum_{i<j} n_i n_j} \sum_{1 \le i < j \le m} \sum_{(o,o^*) \in X_i \times X_j} \mathbb{I}(o^* \succ o) + \frac{1}{2}\mathbb{I}(o^* \sim o),$$

where $X_i \subseteq X$ denotes the set of instances with class labels y_i, and these class labels are assumed to have the order $y_1 < y_2 < \cdots < y_m$. $\mathbb{I}(\cdot)$ is the indicator function mapping false predicates to 0 and true predicates to 1. The training data consists of a set of labeled instances, just like in classification. Since CLeRa is learning from pairwise comparisons of the form (o^*, o), it first extracts such comparisons from the original data by looking at the class information: A preference (o^*, o) is generated for each pair (o^*, y_j) and (o, y_i) of labeled instances in the (original) training data such that $y_i < y_j$.

The ranking performance of CLeRa (with maximum grouping size of $g_{max} = 2$) is compared with LexRank, which was implemented as proposed by (Flach and Matsubara, 2007, 2008); therefore, this method was only applied to binary (two-class) problems but not to problems with more than two classes.[2] We applied naive Bayes (NB) and decision tree (J48) learning as additional baselines, using the standard implementations[3] in the Weka machine learning toolbox Hall et al (2009) and sorting instances according to the estimated probability of the positive class; note that these methods are not applicable to the multi-class case either.

The results of a 10-fold cross-validation are given in Table 1. Since CLeRa produced a completeness of 1 or extremely close to 1 throughout, these values are not reported here. Overall, the performance of the methods is quite comparable but slightly in favour of NB. In particular, CLeRa and LexRank produce quite similar results on many data sets (Asuncion and Newman, 2007). In some cases, however, the results are strongly in favor of CLeRa:

- Census Income: The census data provides information about whether an income exceeds 50,000 USD over a year. The root node of the CLPT is labeled with a single attribute (capital-loss) as well as the descendant node. The preferences on attribute values of the descendant nodes at the third stage depend on the values of the node following the root node. This is also true

[2] The red wine data actually has a target attribute with values between 1 and 10; it was binarized by thresholding at the median.

[3] Trees are not pruned.

Table 1 Average performance in terms of C-index based on a 10-fold cross-validation (best results per data set highlighted in bold font).

Dataset	CLeRa	LexRank	J48	NB
Red Wine	0.7827 ± 0.0479	0.8011 ± 0.0475	0.7378 ± 0.0272	**0.8110** ± 0.0225
Census Income	0.7952 ± 0.0523	0.5776 ± 0.0256	0.7401 ± 0.0356	**0.8607** ± 0.0192
Credit Approval	0.9201 ± 0.0298	**0.9229** ± 0.0389	0.8517 ± 0.0480	0.9061 ± 0.0377
Mammographic Mass	0.8831 ± 0.0289	0.8960 ± 0.0327	0.8524 ± 0.0430	**0.8999** ± 0.0307
Mushroom	**1.0000** ± 0.0000	0.9865 ± 0.0021	**1.0000** ± 0.0000	0.9484 ± 0.0164
SPECT Heart	0.6740 ± 0.0767	0.6590 ± 0.1430	0.5106 ± 0.0961	**0.7409** ± 0.0957
Ionosphere	**0.9198** ± 0.0494	0.5748 ± 0.0740	0.8059 ± 0.1290	0.9061 ± 0.0805
MAGIC Gamma Telescope	0.8218 ± 0.0302	0.7263 ± 0.0517	0.7841 ± 0.0304	**0.8241** ± 0.0329
Breast Cancer Wisconsin	0.9837 ± 0.0171	0.9901 ± 0,0093	0.9793 ± 0.0392	**0.9909** ± 0.0091
German Credit	0.6285 ± 0.0880	0.4523 ± 0.1092	0.6251 ± 0.0902	**0.7835** ± 0.0647
Car Evaluation	**0.9198** ± 0.0185	n/a	n/a	n/a
Nursery	**0.9052** ± 0.0288	n/a	n/a	n/a
Tic-Tac-Toe Endgame	**0.7728** ± 0.0389	n/a	n/a	n/a
Vehicle	**0.7554** ± 0.0459	n/a	n/a	n/a
Cardiocraphic	**0.9551** ± 0.0138	n/a	n/a	n/a

for the importance of the attributes at this stage. One level below, the CLPT also contains nodes that are labeled with grouped attributes.

- Ionosphere: The radar data contains information about whether radar returns are "good" or "bad".[4] With regard to the conditional dependencies and the grouping, the basic structure of the CLPT is very similar to the aforementioned case.
- MAGIC Gamma Telescope: The gamma telescope data contains information about the registration of gamma particles. The basic structure of the CLPT differs from the aforementioned CLPTs with respect to the occurrence of conditional dependencies. Already the first descendant nodes exhibit conditional dependencies on the attribute values of the root node.
- German Credit: In the credit data, customers are classified as good or bad. The respective CLPT makes even stronger use of the proposed extensions compared to the CLPT for the MAGIC Gamma Telescope data set. The first descendant nodes are labeled with grouped attributes.

Overall, these results indicate that the bias imposed by the assumption of a standard lexicographic order is inadequate for these data sets, and hence

[4] Good returns show evidence of some type of structure in the ionosphere.

our extensions (conditional attribute importance, conditional value preferences, variable grouping) clearly pay off.

6 Conclusion and Future Work

Lexicographic orders constitute an interesting model class for preference learning, which allows for representing rankings of a set of objects in a very compact and comprehensible way. Yet, as we have argued in this paper, this model class may not be flexible enough for many real-world applications. Therefore, we have proposed to weaken the assumptions underlying a lexicographic order in various directions, allowing for conditional attribute importance, conditional preferences on attribute values, and variable grouping. Moreover, we have proposed an algorithm called CLeRa, which learns preference models in the form of conditional lexicographic preference trees from training data in the form of pairwise comparisons between objects.

First experimental results in the setting of multipartite ranking are quite promising and show CLeRa to be competitive with other methods. In a direct comparison with an existing lexicographic ranker, the benefit of our extensions are becoming quite obvious.

Important topics of future work can be found both on the theoretical and practical side. In particular, we are currently studying formal properties of our generalized model class, such as its expressiveness and means for regularization and complexity control. Practically, there is certainly scope for improving our current algorithm, for example by devising a suitable procedure for estimating an optimal value g_{max} for the oder of variable grouping. Moreover, improving the computational efficiency of CLeRa would be desirable, too. Last but not least, we are of course interested in real applications for which (generalized) lexicographic models appear to be an adequate representation.

References

Ahlert M (2008) Aggregation of lexicographic orderings. Homo Oeconomicus 25(3):301–317

Asuncion A, Newman DJ (2007) UCI machine learning repository. URL `http://archive.ics.uci.edu/ml/`

Booth R, Chevaleyre Y, Lang J, Mengin J, Sombattheera C (2009) Learning various classes of models of lexicographic orderings. In: Hüllermeier E, Fürnkranz J (eds) Preference Learning, Springer, Berlin, European Conference on Machine Learning and Principles and Practice of Knowledge Discovery in Databases, pp 1–16

Booth R, Chevaleyre Y, Lang J, Mengin J, Sombattheera C (2010) Learning conditionally lexicographic preference relations. In: Proc. ECAI 2010, IOS Press, Amsterdam, The Netherlands, pp 269–274

Boutilier C, Brafman RI, Domshlak C, Hoos HH, Poole D (2004) CP-nets: A tool for representing and reasoning with conditional ceteris paribus preference statements. Journal of Artificial Intelligence 21:135–191

Cheng W, Rademaker M, De Baets B, Hüllermeier E (2010) Predicting partial orders: Ranking with abstention. In: Balcázar J, Bonchi F, Gionis A, Sebag M (eds) Machine Learning and Knowledge Discovery in Databases, Lecture Notes in Computer Science, vol 6321, Springer, Berlin, pp 215–230, DOI 10.1007/978-3-642-15880-3_20

Cohen W, Schapire R, Singer Y (1999) Learning to order things. Journal of Artificial Intelligence Research 10:243–270, DOI 10.1613/jair.587

Dekel O, Manning CD, Singer Y (2003) Log-linear models for label ranking. In: Thrun S, Saul LK, Schölkopf B (eds) Advances in Neural Information Processing Systems, MIT, 16, pp 497–504

Dombi J, Imreh C, Vincze N (2007) Learning lexicographic orders. European Journal of Operational Research 183(2):748–756, DOI 10.1016/j.ejor.2006.10.029

Fishburn PC (1974) Lexicographic orders, utilities and decision rules: A survey. Management Science 20(11):1442–1471, DOI 10.1287/mnsc.20.11.1442

Flach P, Matsubara E (2008) On classification, ranking, and probability estimation. In: de Raedt L, Dietterich T, Getoor L, Kersting K, Muggleton SH (eds) Probabilistic, Logical and Relational Learning - A Further Synthesis, Internationales Begegnungs- und Forschungszentrum für Informatik (IBFI), Schloss Dagstuhl, Dagstuhl, Germany, no. 07161 in Dagstuhl Seminar Proceedings

Flach PA, Matsubara ET (2007) A simple lexicographic ranker and probability estimator. In: Proceedings of the 18th European Conference on Machine Learning, Springer, Berlin, Heidelberg, ECML '07, pp 575–582, DOI 10.1007/978-3-540-74958-5_55

Fürnkranz J, Hüllermeier E (2011) Preference Learning. Springer-Verlag, Berlin, Heidelberg, DOI 10.1007/978-3-642-14125-6

Fürnkranz J, Hüllermeier E, Vanderlooy S (2009) Binary decomposition methods for multipartite ranking. In: Buntine W, Grobelnik M, Mladenić D, Shawe-Taylor J (eds) Machine Learning and Knowledge Discovery in Databases, Lecture Notes in Computer Science, vol 5781, Springer, Berlin, pp 359–374, DOI 10.1007/978-3-642-04180-8_41

Hall M, Frank E, Holmes G, Pfahringer B, Reutemann P, Witten IH (2009) The WEKA data mining software: An update. SIGKDD Explorations 11(1):10–18, DOI 10.1145/1656274.1656278

Hüllermeier E, Fürnkranz J, Cheng W, Brinker K (2008) Label ranking by learning pairwise preferences. Artificial Intelligence 172(16–17):1897–1916, DOI 10.1016/j.artint.2008.08.002

Schmitt M, Martignon L (2006) On the complexity of learning lexicographic strategies. Journal of Machine Learning Research 7:55–83

Wilson N (2009) Efficient inference for expressive comparative preference languages. In: Proceedings of the 21st International Joint Conference on Artificial Intelligence, Morgan Kaufmann, San Francisco, IJCAI'09, pp 961–966

Yaman F, Walsh TJ, Littman ML, desJardins M (2008) Democratic approximation of lexicographic preference models. In: Proc. ICML-08, Helsinki, Finland, pp 1200–1207

Application of Classification Trees in the Analysis of the Population Ageing Process

Justyna Wilk

Abstract A process of socio-economic development is continuously accompanied by a process of population ageing. In terms of a policy of regional development, it is valuable to identify factors of ageing to mitigate or impede its undesirable impact on a national economy. The paper discusses how to model the process of ageing using classification trees and presents an empirical study. The main research question is if the populations similar in their degrees of ageing, feature common demographic conditions of this process as well.

1 Introduction

The socio-economic development process is inseparably accompanied by the population ageing process (see Uhlenberg, 2009; Martinez-Fernandez et al, 2012). Population ageing relates to the changes in the age distribution and an increase in the percentage share of senior citizens in the general population. This results from reductions in mortality which are followed by reductions in the number of births. It is mostly determined by social, economic, cultural, environmental and other factors, such as: the intentional delay of procreation

Justyna Wilk
Wrocław University of Economics, Department of Econometrics and Computer Science, Nowowiejska 3 Street, 58-500 Jelenia Góra, Poland,
✉ justyna.wilk@ue.wroc.pl

ARCHIVES OF DATA SCIENCE, SERIES A
KIT SCIENTIFIC PUBLISHING
Vol. 1, No. 1, S. 57–76, 2016

DOI 10.5445/KSP/1000058747/04
ISSN 2363-9881

time and changing life priorities, leading more and more towards a healthy lifestyle, progress in medicine etc.

Population ageing strongly affects the situation of a country in terms of its financial, social and economic conditions. In the long-term prospect, without taking appropriate actions, it may lead to disturbing the retirement system, decreasing the efficiency of social systems and an increasing gap in the labour market etc. (see Magnus, 2008; Weil, 1997; United Nations, 2013; Légaré, 2006).

Population ageing is a natural process in economically well developed countries (see Prskawetz and Lindh, 2011; Lindh and Malmberg, 2009; Martinez-Fernandez et al, 2012). The most intensive process of population ageing is seen in Asian countries, especially in Japan, where one in four people is 65 years old or older (see World Population Prospects, 2012). This process also occurs in the majority of European Union (EU) countries of which Italy and Germany are demographically the oldest. Although the population of countries which joined the EU in 2004 and 2007 is relatively young (Slovakia, Poland and Cyprus are the youngest), they also suffer from ageing (see Muenz, 2007; Giannakouris, 2008; Grundy, 1996).

The diversification of demographic situations relates not only to the national economies but also occurs within countries. This results from socio-economic disparities and also cultural, social and environmental differences etc. Martinez-Fernandez et al (2012) see ageing as one of the most substantial factors of global demographic change and the shrinkage of cities and regions as well.

Regions differ in their nature of ageing; its intensity and demographic conditions, e.g. extremely low fertility, high out-migrations of young people, older and older workforce etc. Different situations of regions require different activities to mitigate the effects of ageing and also to impede this process. The identification of factors of ageing may be helpful in creating a policy of regional development.

This paper proposes using classification trees to model the population ageing process and presents an empirical study. The key research question is if regions similar in their intensity of ageing, feature common factors of this process as well.

2 Modelling the population ageing process

An examination of ageing covers its intensity and the demographic determinants. A lot of research studies are limited to examining the degree of ageing. This requires defining the threshold of ageing and determining the age structure of a population.

The threshold of ageing is the age which classifies a person as being older. Some statistical studies use the retirement age as the threshold of ageing, which is usually separate for men and women. Other research studies assume the age of 60 or 65 as threshold and the same threshold for both, men and women (see e.g. Sauvy, 1948). Sanderson and Scherbov (2008) see the age for which the remaining life expectancy is 15 years as the threshold of ageing.

Several researchers propose a lot of rates indicating the age structure of a population (see Clarke, 1965; Beaujeu-Garnier, 1966; United Nations, 2013). The first group includes age-related measures e.g. measures of location (e.g. median age), the old-age rate, ageing index, old-age dependency ratio. The second group covers the graphical tools such as, the population pyramid and Ossan's triangle etc.

Population ageing results from reductions in mortality and the number of births. At the national level, fertility and life expectancy are seen as the main determinants of the age structure (see Preston et al, 1989; United Nations, 2013). At the regional level, it may also be affected by massive out-migration of young people. The process of ageing is also intensified by the ageing of the adult population, if the share of older adult people is increasing. Moreover, the advancement of ageing is seen, in a population with a very high proportion of people older than 80 years, in the population aged 60 and older.

The study of ageing requires examination if regional sub-populations with similar intensity of ageing also feature common demographic conditions of this process. We can use a contingency table which displays the multivariate frequency distribution. This approach is simple for categorical data or if a division of numerical data is given.

Otherwise we can use multivariate data analysis (see Colley and Lohnes, 1971; Johnson and Wichern, 1992; Everitt and Dunn, 2001; Hair et al, 2006). A regression analysis can identify an overall impact of the demographic factors on the intensity of population ageing and helps to select variables for further studies. But in terms of creating a policy of regional development, it is valuable to identify common factors of ageing in sub-populations exhibiting a similar intensity of this process.

Discriminant analysis is useful when simple, linear interactions between a dependent and the explanatory variables exist, and when the random variables in the model follow a multivariate normal distribution (see Fisher, 1936; Klecka, 1980; Lachenbruch, 1975). In contrast to this, classification trees are nonparametric and the assumption of a multivariate normal distribution is not relevant. They are useful, when complicated and nonlinear relations exist (see Breiman et al, 1984; Ripley, 1996; Rokach and Maimon, 2008).

Classification trees are widely used in medicine, computer science, botany, psychology, finance, marketing, engineering etc. rather than in demography. But Ninčević et al (2010) used them to examine the impact of various factors on life expectancy. Toulemon (2006) applied them to model the transition to adulthood and find differences between Austrians' and Italians' paths. High predictive performance, simplicity and transparency of the classification scheme make them a promising tool in examining the population ageing as well.

3 Classification trees in the analysis of population ageing

This paper examines the situation of Poland which is one of the fastest developing EU countries. Although the Polish population is relatively young, it exhibits an intensive demographic ageing process. The identification of the degree of ageing and its regional diversification, and finding its demographic factors is crucial in order to impede or mitigate the effects of ageing and program the right regional development policy. The main interest of the study is to examine if the sub-populations similar in their degrees of ageing, feature common demographic conditions of this process as well.

The investigation was carried out for 66 Polish subregions. These NUTS 3 units are statistical rather than administrative territorial units. They come from a division of 16 Polish provinces. As Polish subregions represent relatively homogeneous territorial areas in respect to economic, social, cultural and environmental features, we can assume that they also exhibit individual demographic situations. The study covers the period 1995-2012 because the demographic changes can only be seen over a long time period. Empirical data comes from the Central Statistical Office of Poland.

This paper uses classification trees in the comparative analysis of demographic conditions of population ageing. Classification trees are used to predict a membership of units in the categories of a categorical dependent (response) variable from their measurements on a set of predictors (explanatory variables). In this study, they help to find differences between predefined classes of subregions and identify profiles of these classes. The next sections follow a classification procedure:

1. Specifying the demographic features of population ageing. Sect. 4 identifies the degrees of ageing in Polish subregions, while Sect. 5 discusses the demographic factors affecting this process.
2. Classifying a set of subregions. Sect. 6 describes the construction and estimation of a classification tree and presents the classification results.
3. Profiling classes. Sect. 7 identifies demographic conditions of population ageing and makes policy suggestions. Sect. 8 gives comments and open questions.

4 The intensity of the population ageing process

The comparative studies of demographic ageing usually use a typology of the population age. Each of them takes some indicator of a population structure and distinguishes a set of categories of demographic age according to a value of this indicator. The first typology was developed by Sündbarg (1900), the others were proposed by Sauvy (1948), Beaujeu-Garnier (1966), and Veyret-Verner (1971) etc.

In their recent works, the United Nations (UN) distinguishes five demographic ages according to the share of senior citizens in the total population. If the percentage of people aged 65 or older is less than 4%, the population is demographically young. A share above 4% and below 7% indicates a mature population. A share below 14% means an ageing population while a share between 14% and 21% defines an old population. More than 21% is typical of the aged population (see Coulmas, 2007, p. 5).

The population age defines the present demographic situation of a population but does not indicate the expected changes in the age structure. Therefore, in analyzing a relatively young population such as Poland, we should also consider the increase rate of the share of senior citizens. A value of the rate shows a scale of changes and can be used to foresee the future age structure.

Table 1 Basic statistics of the share of people aged 65 or older in 2012, and the increase rate of this share in the period from 1995 to 2012 in 66 Polish subregions (NUTS 3 units)

Variable name	Share of senior citizens in 2012 (%)	Increase rate of the share in the years 1995-2012 (%)
Poland	14.2	1.4
Minimum	9.9	0.2
Maximum	18.6	3.4
Median	13.9	1.4
Mean	14.1	1.5
Standard deviation	1.7	0.7
Coefficient of variation (%)	12.2	46.6
Pearson's correlation $[-1,1]$	0.08	

Over the last thirty years, the share of young people in Poland has significantly decreased, while the number of older people increased. The yearly increase of the share of senior citizens was 1.4% from 1975 to 2012 to reach 14.2% of people aged 65 or older in 2012 (see Table 1).

This situation is not homogeneous in the whole country. High socio-economic disparities and also cultural, social and environmental differences translate into a territorial diversification of the age structure and its changes in time. The values of the share of senior citizens were in the range of [9.9, 18.6] in 66 Polish subregions in 2012 (see Table 5 for details). According to the UN classification, we can distinguish subregions with ageing (53%) and subregions with old (47%) population, respectively.

All subregions showed yearly increases of this share; the majority of them showed a yearly increase between 0.5 and 2.0%. According to the national average value (1.5%), two groups of subregions were distinguished: relatively low (56% of subregions) and high progress of ageing (44% of subregions).

It is interesting that the share of senior citizens is not statistically correlated with the increase rate of this share. We cannot conclude that the higher the increase rate of people aged 65 and older in the total population, the older the population (and inversely). It is also not true that the older the population, the lower the increase rate of the share (and inversely). We can use both features, the demographical age of a population and the progress of ageing to divide subregions into four classes:

1. The "old-high" class includes 24.2% of subregions with a high share of senior citizens and also a high growth rate of this share.

2. The "old-low" class consists of 22.7% of subregions with a high share of senior citizens but with a low growth rate of this share.
3. The "ageing-high" class is represented by 19.7% of subregions with a low share of senior citizens but with a high growth rate of this share.
4. The "ageing-low" class is the biggest and covers 33.3% of subregions with a low share of senior citizens and also with a low growth rate of this share.

In general, the most intensive ageing relates to the majority of subregions located in south western Poland (see Fig. 1). This process is also seen in the biggest Polish cities. Demographically old, but not ageing quickly, is the population of east Poland and also some subregions of central Poland which belong to the Łódzkie and Świętokrzyskie provinces). Still relatively young but quickly ageing is the population of north western and west Poland and also the Warmińsko-mazurskie province.

5 Factors of the population ageing process

This section examines five demographic factors of ageing (discussed in Sect. 2) relevant for the Polish population. An indicator of the generation replacement is the total fertility rate. It is the average number of children a woman would bear over the course of her lifetime if current age-specific fertility rates remained constant throughout her childbearing years. Values between 2.10 and 2.15 provide the replacement of a generation without excessive growth or shrinkage of a population.

The rate reached approximately 1.30 in Poland and from 0.9 to 1.6 in subregions but 85% of them did not reach 1.4 in 2012 (see Table 2). Although the replacement condition was not satisfied, the fertility rate has been increasing. But this growth is extremely slow and we cannot expect any extra changes in the next 20 years.

An indication of life duration is the life expectancy of people at the age of 65. It is the mean number of years still to be lived by a man or woman who has reached the age of 65, if subject throughout the rest of his/her life to the current mortality conditions (age-specific probabilities of dying). The Polish population is living longer and longer. Yearly increases in the life expectancy of men and women at the age 65 are observed in all subregions. An average 65 year old man will live to be 80.4 years old which is approximately 4.0 years less than an average woman. The statistical correlation of the indicator values

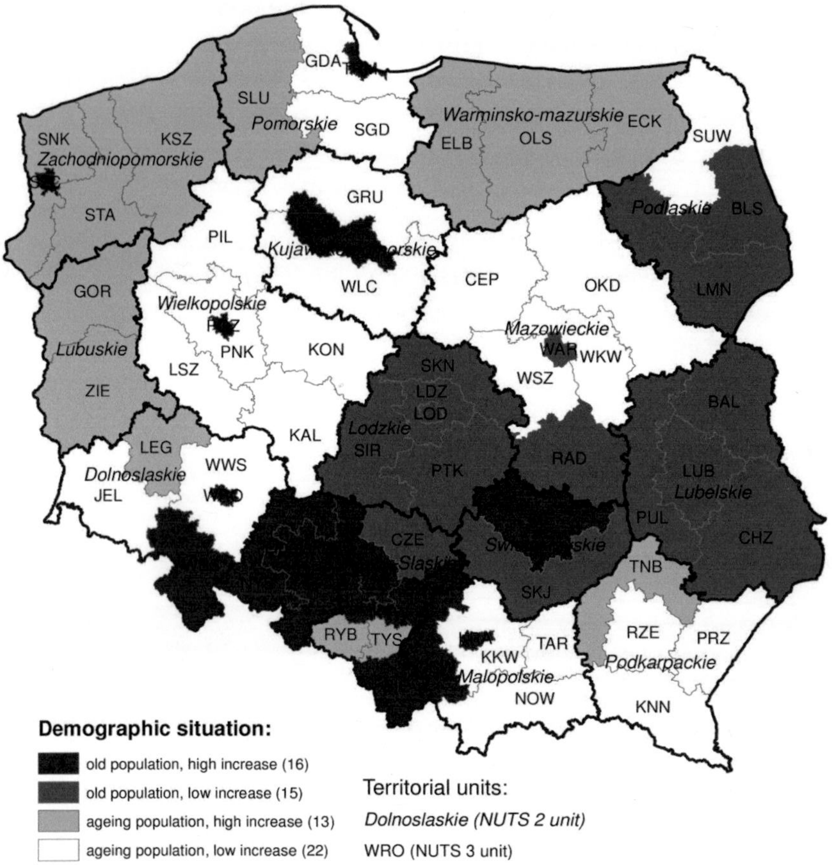

Fig. 1 Four classes of 66 Polish subregions (NUTS 3 units) according to the demographic age in 2012 and the progress of population ageing in the years 1995-2012 (see Table 5 for details)

for men and women in subregions is very high. Thus the study includes only the situation of men.

An indicator of changes in a demographic structure is the net migration of people in the age between 20 and 59. It is the difference between registered (domestic and international) migration inflow and outflow of people with a permanent residence in a region to the average population of the region. The rate took values between –55 and +100 in 2012 in Polish subregions. In 2012 nearly 75% of subregions had a negative balance of migration (higher outflow than inflow of people). Yearly increases of the rate were seen in the majority of

Table 2 The demographic factors of population ageing in Polish subregions (NUTS 3 units)

Abbreviation	Variable name	Year*	Poland	Median	Minimum	Maximum
Fertility	Total fertility rate (person)	2002	1.249	1.312	0.893	1.614
		2012	1.299	1.298	1.091	1.632
Life	Life expectancy of men at the age 65 (year)	2007	14.6	14.3	13.6	16.4
		2012	15.4	15.2	14.2	16.9
Migration	Net migration rate of population aged 20-59	1995	*NA*	*NA*	*NA*	*NA*
		2012	*NA*	–21.2	–55.2	100.2
Working	Working-age population ageing rate (%)	2009	37.8	37.7	33.7	41.7
		2012	37.4	37.7	34.4	41.1
Oldest	Oldest-old-age population rate (%)	2005	20.3	20.3	16.2	24.2
		2012	26.3	26.3	19.5	31.3

*depending on the data availability, *NA* – not applicable.

subregions. A much higher inflow than outflow of young people is seen in the biggest Polish cities and their surroundings.

An indicator of ageing of an adult population is working-age population ageing rate. The working age is the age between 18 and the retirement age. The age 45 distinguishes the immobile and mobile working age. The working-age population ageing rate means the percentage of immobile working-age people in the working-age population. The rate was 37.4% in Poland in 2012 but four subregions exceeded the threshold of 40.0%.

The progress of ageing of the senior population is shown by the oldest-old-age population rate. It is the percentage of people aged 80 and older in the total population aged 65 and older. The rate reached 26.3% in Poland in 2012. That was 6 percentage points more than in 2005. But two subregions exceeded 30% while one subregion did not reach 20.0%. Yearly increases of the rate were seen in all subregions.

Two separate linear regression models served to examine the statistical significance of these demographic factors in explaining the share of senior citizens (see Table 3) and its increase rate (see Table 4). The estimate of the total fertility rate is statistically significant exclusively for the increase rate which means that the impact of fertility is perceptible in the long term rather than in the short term. The higher the fertility, the smaller the progress of ageing. A Similar situation relates to the net migration rate whose estimate is significant only for the increase rate of the share of seniors. The lower the net migration rate, the older the population.

Both estimates of life expectancy are significant and positive. The increase of life expectancy in a population causes an increase of the share of senior citizens more intensively than the increase rate of this share. A similar situation is presented by the working-age population ageing rate. The oldest-old-age population rate is also related to both dependent variables but proved bilateral impact. The older the senior population, the older the whole population but the lower the progress of ageing.

6 Construction and estimation of a classification tree

Poland has high regional disparities according to the share of senior citizens and its increase rate. Four classes of subregions presenting different degrees of ageing were distinguished in Sect. 4. Although all demographic factors affecting the situation of subregions discussed in Sect. 5 are statistically significant, there are no obvious differences between classes according to the average values of demographic factors (see Table 5). For example, the averages of the working-age population ageing rate are very similar in old-high, old-low and ageing-high classes.

We cannot profile classes on the basis of these results. In this situation we construct a classification tree to depict relations between the degrees of population ageing and a set of demographic factors of this process. The set of four classes of subregions served as the realizations of categorical dependent variable, while all demographic factors formed a set of explanatory variables.

In the study, the CART algorithm proposed by Breiman et al (1984) is applied to profile pre-assigned classes for subregions. CART produces a tree-structured model using recursive binary partitioning. The algorithm asks a sequence of questions which split a set of objects into two subsets. Splitting is determined by a condition of the value of a single explanatory variable which is satisfied by the observation or it is not.

The starting point of a tree, called a root, consists of the entire learning set. A set of nodes originates from the root. A nonterminal (or parent or internal) node is a node that splits into two daughter nodes. A node which has stopped splitting is called a terminal node (or a leaf). A path from the root to the terminal node shows the classification rules on which the units are assigned to a class.

The construction and estimation of a tree includes three steps: the partitioning of the dataset, determining the complexity of the tree and validating the results.

Table 3 Estimation results for the share of senior citizens in 2012 (least-squares method)

Name*	Coefficient	Standard error	Student's t-test	p-value
Constant	–52.664	7.196	–7.318	6.05e-010[a]
Fertility	x	x	x	x
Life	2.005	0.285	7.034	1.87e-09[a]
Migration	x	x	x	x
Working	0.821	0.108	7.601	1.95e-010[a]
Oldest	0.205	0.066	3.086	0.003[a]
R-Squared		0.604778		

* Fertility: the total fertility rate (person), Life: the life expectancy of men at the age 65 (year), Migration: the net migration rate of population aged 20-59, Working: the working-age population ageing rate (%), Oldest: the oldest-old-age population rate (%)
[a] denotes statistical significance at the 99% level
x denotes statistical insignificance

Table 4 Estimation results for the increase rate of the share of senior citizens in the years 1995-2012 (least-squares method)

Name*	Coefficient	Standard error	Student's t-test	p-value
Constant	0.855	3.714	0.230	0.8187[b]
Fertility	–1.981	0.546	–3.626	0.0006[a]
Life	0.452	0.111	4.080	0.0001[a]
Migration	–0.004	0.001	–3.914	0.0002[a]
Working	0.084	0.045	1.869	0.0665[b]
Oldest	–0.265	0.019	–14.260	6.27e-021[a]
R-Squared		0.817918		

* Fertility: the total fertility rate (person), Life: the life expectancy of men at the age 65 (year), Migration: the net migration rate of population aged 20-59, Working: the working-age population ageing rate (%), Oldest: the oldest-old-age population rate (%)
[a] and [b] denote statistical significance at the 99% and 90% levels respectively

Table 5 The average values of demographic factors determined for four classes of regions in 2012

Variable name*	Old-high class	Old-low class	Ageing-high class	Ageing-low class
Fertility	1.199	1.289	1.288	1.379
Life	15.5	15.2	14.9	15.3
Migration	–7.6	–16.2	–21.4	3.5
Working	38.1	37.9	38.1	36.5
Oldest	25.3	28.1	24.6	26.1

* Fertility: the total fertility rate (person), Life: the life expectancy of men at the age 65 (year), Migration: the net migration rate of population aged 20-59, Working: the working-age population ageing rate (%), Oldest: the oldest-old-age population rate (%)

The tree growing procedure is based on the partitioning rules. They divide subsets of the learning set with respect to a dependent variable and create the daughter nodes from a parent node. The data in each of the daughter nodes is obtained by reducing the number of cases that has been misclassified. All of the possible ways of splitting are tested and the one which leads to the greatest increase in node purity is chosen. The goodness of a potential split is indicated by an impurity function. This is a function of the proportion of the learning sample belonging to the possible classes of the dependent variable. This study uses Gini's index of impurity.

The second problem of classification is to select a tree of the right size in such a way as not to overfit the learning sample, as well as to achieve an exhaustive representation of the data. In this study, the dataset is relatively small, so we use a strategy of growing a fully expanded tree and then pruning it back, or removing some of its nodes to produce a tree with a smaller number of terminal nodes. This produced a finite sequence of nested subtrees from which the best solution was chosen.

The membership accuracies in a sequence of subtrees can be compared using some estimates of their misclassification rates. One of the possible solutions, if one has enough data, is to distinguish an independent test set. In this study, due to having a not very large number of cases, a V-fold cross-validation estimate of the misclassification rate is used. The entire dataset used as a learning set is randomly divided into V parts of approximately equal sized, disjoint subsets. In this study V was equal to 5.

The subtree with the smallest estimated misclassification rate equal to 12.1% is selected to be the final tree-based classification model. Table 6 presents a tree growing scheme. The classification tree has 10 internal nodes and 12 leaves which determine the profiles of classes. The node numbers indicate the rules of a division. For example, the leaf node with the number of 2.2.2.2.2.1 is a subdivision of an internal node with the number of 2.2.2.2.2 according to the values of the life expectancy of men at the age 65 higher than 14.85 years. All calculations were made in Statistica 10 software.

An example of the classification rules for a subclass of the ageing-low class is the leaf node number 2.2.2.1.2 in Table 6. This subclass includes subregions with the rate of adult population ageing between 37.09% and 37.78% which is very close to the Polish national average equal to 37.4% in 2012. The percentage of people aged 80 and older in the total population is lower than 27.60%. The life expectancy of men at the age 65 is less than 15.15 years which is rather shorter than the Polish national average equal to 15.4% in 2012. The total

Table 6 Classification tree

Node number	Type of node	Input variable*	Splitting criterion	Number of sub-regions (66)	Number of old-high class members (16)**	Number of old-low class members (15)**	Number of ageing-high class members (13)**	Number of ageing-low class members (22)**
1	internal	Working	≤37.09	26	3	4	1	**18**
1.1	leaf	Fertility	>1.25	18	0	1	0	**17**
1.2	internal	Fertility	≤1.25	8	**3**	3	1	1
1.2.1	leaf	Working	≤35.99	3	**3**	0	0	0
1.2.2	internal	Working	>35.99	5	0	**3**	1	1
1.2.2.1	leaf	Migration	≤−18.18	2	0	0	1	**1**
1.2.2.2	leaf	Migration	>−18.18	3	0	**3**	0	0
2	internal	Working	>37.09	40	**13**	11	12	4
2.1	leaf	Oldest	>27.60	6	0	**6**	0	0
2.2	internal	Oldest	≤27.60	34	**13**	5	12	4
2.2.1	internal	Life	>15.15	9	**8**	0	1	0
2.2.1.1	leaf	Life	>15.30	6	**6**	0	0	0
2.2.1.2	leaf	Life	≤15.30	3	**2**	0	1	0
2.2.2	internal	Life	≤15.15	25	5	5	**11**	4
2.2.2.1	internal	Fertility	>1.37	5	0	**3**	0	2
2.2.2.1.1	leaf	Working	>37.78	3	0	**3**	0	0
2.2.2.1.2	leaf	Working	≤37.78	2	0	0	0	**2**
2.2.2.2	internal	Fertility	≤1.37	20	5	2	**11**	2
2.2.2.2.1	leaf	Fertility	>1.28	11	0	1	**9**	1
2.2.2.2.2	internal	Fertility	≤1.28	9	**5**	1	2	1
2.2.2.2.2.1	leaf	Life	>14.85	3	0	**1**	1	1
2.2.2.2.2.2	leaf	Life	≤14.85	6	**5**	0	1	0

* Fertility: the total fertility rate (person), Life: the life expectancy of men at the age 65 (year), Migration: the net migration rate of population aged 20-59, Working: the working-age population ageing rate (%), Oldest: the oldest-old-age population rate (%)
** Bold values indicate a predicted class for which a cost of misclassification is the lowest

fertility rate is higher than 1.37 which is much more than the Polish national average equal to 1.299 in 2012. A population of subregions belonging to this subclass is relatively young and does not age very quickly due to presenting a relatively short life duration and high fertility as well.

7 Demographic conditions of population ageing within Poland

The results of a classification tree help to profile 4 classes of subregions. The first (old-high) class with an old and still ageing population consists of 3 subclasses. The most intensive ageing is seen in the big socio-economic centers in Poland:

the cities of Cracow (KRA), Wrocław (WRO) and Poznań (POZ). See Table 7. The GDP *per capita* in these cities is 150% of the national average value (Poland: 37,096 PLN). In spite of the relatively young working-age population, extremely low fertility accompanied by high life expectancy is seen in these subregions.

This probably results from the intentional delay of procreation time and changing life priorities from family-related to profession-related. The situation requires a redefinition of social policy. Exemplary activities include promoting starting a family, giving help to women to reconcile their career with caring for children, e.g. by providing social infrastructure (e.g. increasing the availability of nurseries), giving an opportunity for men to take paternal leave etc.

The second subclass covers economically well developed subregions with the cities inside and also the city of Szczecin (SZC). They exhibit an extremely old working-age population which is accompanied by low fertility and high life expectancy. These subregions experience serious demographic problems which are difficult to solve. The population of young people is shrinking. This can disturb the regional labour market and diminish demand for jobs. Regional policy, e.g. creating new jobs and services, should attract people to live in these subregions.

The intensive process of population ageing is also typical of weak, neighbouring subregions of south-western Poland apart from the Katowicki (KTW) subregion. Extremely old working-age population and low fertility, and also a big migration outflow of young people occur in this area. These subregions suffer from very serious demographic and economic problems which affect a depopulation process and may lead these subregions to become extinct.

The second (old-low) class with an old but very slowly ageing population also distinguishes three subclasses. The first subclass covers a few subregions together with Warsaw (WAR), the capital city of Poland. The situation of this subclass is very similar to the situation of the cities of Cracow, Wrocław and Poznań. But it will be much more difficult to increase fertility in Warsaw due to social and cultural (e.g. life style) and economic (e.g. high cost of living) adversities.

The old-low class also includes the weakest Polish subregions (less than 75% of the national average value of GDP *per capita* value in 2012) with common borders such as the Chełmsko-zamojski (CHZ), Puławski (PUL), Radomski (RAD) and Sandomiersko-jędrzejowski (SKJ) subregions. Extremely high migration outflow of young people from these subregions and also a lot of oldest-old age people directly result from the long-term economic problems.

Table 7 Share of people aged 65 or older and its increase rate in 66 Polish subregions (NUTS 3)

Name	Full name	Share of of senior citizens in 2012 (%)	Increase rate of the share in 1995-2012 (%)	Name	Full name	Share of of senior citizens in 2012 (%)	Increase rate of the share in 1995-2012 (%)
BAL	Bialski	14.1	0.4	OPO	Opolski	15.0	2.8
BIE	Bielski	14.4	1.7	OSW	Oświęcimski	14.5	1.8
BLS	Białostocki	14.6	1.3	PIL	Pilski	11.9	1.3
BOT	Bydgosko-toruński	14.3	1.9	PNK	Poznański	10.7	0.5
BYM	Bytomski	15.4	2.8	POZ	City of Poznań	16.4	1.6
CEP	Ciechanowskopłocki	13.9	1.1	PRZ	Przemyski	13.9	0.9
CHZ	Chełmsko-zamojski	15.3	0.7	PTK	Piotrkowski	14.3	0.9
CZE	Częstochowski	15.8	1.3	PUL	Puławski	15.6	1.0
ECK	Ełcki	12.3	1.9	RAD	Radomski	14.1	0.9
ELB	Elbląski	12.2	1.6	RYB	Rybnicki	13.6	3.4
GDA	Gdański	9.9	1.4	RZE	Rzeszowski	13.6	1.2
GLI	Gliwicki	15.3	3.2	SGD	Starogardzki	11.2	1.4
GOR	Gorzowski	12.5	1.6	SIR	Sieradzki	15.0	0.8
GRU	Grudziądzki	12.6	1.1	SKJ	Sandomiersko-jędrzejowski	16.1	0.3
JEL	Jeleniogórski	13.8	1.3	SKN	Skierniewicki	15.5	1.0
KAL	Kaliski	13.5	1.0	SLU	Słupski	11.9	1.9
KIE	Kielecki	15.3	1.8	SNK	Szczeciński	11.4	1.9
KKW	Krakowski	13.0	0.2	SOS	Sosnowiecki	15.7	1.9
KNN	Krośnieński	13.7	1.4	STA	Stargardzki	12.3	1.6
KON	Koniński	13.0	1.3	SUW	Suwalski	13.9	1.0
KRA	City of Kraków	16.6	1.9	SZC	City of Szczecin	16.0	2.2
KSZ	Koszaliński	13.0	2.1	TAR	Tarnowski	13.9	1.5
KTW	Katowicki	16.2	2.7	TNB	Tarnobrzeski	13.4	1.8
LDZ	Łódzki	15.5	1.1	TRM	Trójmiejski	17.0	2.3
LEG	Legnicko-głogowski	12.7	2.5	TYS	Tyski	12.4	2.7
LMN	Łomżyński	16.2	0.9	WAR	City of Warsaw	18.0	1.2
LOD	City of Łódź	18.6	1.2	WBS	Wałbrzyski	15.2	1.5
LSZ	Leszczyński	12.1	0.8	WKW	Warszawski wschodni	12.8	0.6
LUB	Lubelski	14.6	1.3	WLC	Włocławski	13.4	1.2
NOW	Nowosądecki	12.5	1.4	WRO	City of Wrocław	16.6	1.7
NYS	Nyski	14.4	1.7	WSZ	Warszawski zachodni	13.6	0.5
OKD	Ostrołęcko-siedlecki	13.9	0.5	WWS	Wrocławski	12.1	0.8
OLS	Olsztyński	12.5	2.0	ZIE	Zielonogórski	12.8	1.5

This is a very similar situation to the last subclass of the old-low class. An effective social policy and economic measures to stimulate regional economies may help to keep young people and impede shrinkage.

The third (ageing-high) class is formed by subregions with a relatively young population which experiences the ageing process. It is the most homogeneous class according to demographical conditions but differs in economic and environmental factors. This class consists of selected subregions of the Lubuskie region (the Gorzowski (GOR) and Zielonogórski (ZIE) subregions) located in western Poland, subregions of the Śląskie region (the Rybnicki (RYB) and Tyski (TYS) subregions) located in southern Poland, the Warmińsko-mazurskie region (the Elbląski (ELB) and Ełcki (ECK) subregions) located in northern Poland, and also the Słupski (SLU) and Legnicko-głogowski (LEG) subregions. Low fertility, accompanied by an old working-age population, which results from the migration outflow of young people, are typical features of this class. Some of these subregions will probably join the old-high class in the future.

The fourth (ageing-low) class presents the smallest population ageing but is the most diversified according to its demographic conditions. The majority of its subregions has a relatively young working-age population but also high life expectancy. Extremely weak economic situation (GDP *per capita* less than 60% of the national average value) translates into a high migration outflow of young people.

Subregions surrounding the biggest economic centers of Poland (the cities of Warsaw, Wrocław, Poznań, Cracow and Gdańsk) show an extremely high positive net migration of young people. This results from the sub-urbanization process and expansion of these cities. The rest of the subregions of the ageing-low class home a relatively old working-age population and low fertility, accompanied by a high migration outflow of young people. This subclass will probably join the ageing-high class if no preventive measures are taken.

8 Discussion

Population ageing is not simple to examine because its foundations do not lie only in demographic factors. For example, in general, we can assume that subregions similar in their economic situation are also similar demographically. However, when we compare the situation of the biggest Polish cities with the other economically well developed subregions, we cannot expect that the demographic situation of such different types of agglomerations will be similar.

The occurring sub-urbanization process may also disturb the analysis results. Thus we cannot expect that the youngest population lives in the biggest cities.

Many young people work in a city but live outside the city. It is a problem how to interpret this situation. Is the population of economic centres really demographically old?

Some demographic factors of the population ageing process which were not disclosed in this study also exist. For example, eastern Poland is economically the weakest macro-region, so we can suppose that the population of the subregions of eastern Poland is demographically the oldest. But there are subregions in the rest of Poland which are economically well developed and also show a very intensive process of population ageing. For example, some pensioners, after working many years outside, come back to their native region, e.g. to the Śląskie region, to spend the rest of their lives there.

Although demographic changes are not rapid and take a long-time, the analysis shows that results are not stable in time. Changes of economic trends, social and economic policies, actions of local authorities, and even some misfortunes affect the demographic situation. Carrying out additional studies is recommended.

9 Conclusions

The paper conducts a comparative study of demographic features of populations exhibiting the ageing process using classification trees. It presents an original approach in the analysis of population ageing which is not well recognized in the literature.

1. It examines the demographic determinants of population ageing besides the intensity of ageing.
2. It concerns the internal situation of a country rather than international comparisons.
3. It uses classification trees which are not frequently applied in demography and population ageing studies.
4. It indicates regional problems and conditions of ageing in Poland.

The main advantage of using classification trees is to discover two, previously unknown situations difficult to find with other tools. The most important result of the study is to prove that regions similar in their intensity of population ageing may feature different demographic conditions of the process. Only subregions with young but quickly ageing population are common in their factors of ageing.

The second, unexpected finding is that the intensity of population ageing is not correlated with the economic situation of subregions. But subregions similar in their economic situation exhibit common demographic factors of ageing as well. In general, well developed regions mostly suffer from very low fertility, while poor regions struggle with a high out-migration of young people and an old working-age population. This requires an individual approach in mitigating the effects of ageing and impeding this process within a country and, besides an effective national social policy and also coordinated activities of regional and local authorities.

References

Beaujeu-Garnier J (1966) Geography of population. Longmans, London

Breiman L, Friedman JH, Olshen RA, Stone CJ (1984) Classification and Regression Trees. Statistics/Probability Series, Wadsworth, Belmont, California

Clarke JI (1965) Population geography. Pergamon Press, Oxford

Colley WW, Lohnes PR (1971) Multivariate data analysis. Wiley, New Jersey

Coulmas F (2007) Population decline and Ageing in Japan – the social consequences. Routledge, New York

Everitt BS, Dunn G (2001) Applied Multivariate Data Analysis. Wiley, London

Fisher RA (1936) The use of multiple measurements in taxonomic problems. Annals of Eugenics 7(2):179–188, URL http://hdl.handle.net/2440/15227

Giannakouris K (2008) Aging characterises the demographic perspectives of the european societies. Eurostat Statistics in Focus 72:1–11

Grundy E (1996) Population ageing in europe. In: Coleman D (ed) Europe's Population in the 1990s, Oxford University Press, Oxford, pp 267–299

Hair JF, Black WC, Babin BJ, Anderson RE, Tatham RL (2006) Multivariate Data Analysis, 6th edn. Prentice Hall, New Jersey

Johnson RA, Wichern DW (1992) Applied multivariate statistical analysis, 3rd edn. Prentice-Hall, New Jersey

Klecka WR (1980) Discriminant analysis. Sage Publications, Beverly Hills

Lachenbruch PA (1975) Discriminant analysis. Hafner Press, New York

Légaré J (2006) Economic, social and cultural consequences of the ageing of the population. In: Caselli D, Vallin J, Wunsch G (eds) Demography - Analysis and Synthesis: A Treatise in Population, Elsevier, London, pp 327–336

Lindh T, Malmberg B (2009) European union economic growth and the age structure of the population. Economic Change and Restructuring 42(3):159–187, DOI 10.1007/s10644-008-9057-1

Magnus G (2008) The age of aging: how demographics are changing the global economy and our world. Wiley, New York

Martinez-Fernandez C, Kubo N, Noya A, Weyman T (2012) Demographic Change and Local Development: Shrinkage, Regeneration and Social Dynamics. OECD Publishing, Paris, DOI 10.1787/9789264180468-en

Muenz R (2007) Aging and demographic change in european societies: Main trends and alternative policy options. SP Discussion Paper 703, World Bank Group

Ninčević I, Ćukušić M, Garača Z (2010) Mining demographic data with decision trees. In: MIPRO, Proceedings of the 33rd International Convention, IEEE, Opatija, Croatia, pp 1288–1293

Preston SH, Himes C, Eggers M (1989) Demographic conditions responsible for population ageing. Demography 26(4):691–704, DOI 10.2307/2061266

Prskawetz A, Lindh T (2011) The relationship between demographic change and economic growth in the EU. Austrian Academy of Sciences Publishing House, Wien

Ripley BD (1996) Pattern recognition and neural networks. Cambridge University Press, Cambridge

Rokach L, Maimon O (2008) Data mining with decision trees: theory and applications. World Scientific, New Jersey Publishing

Sanderson W, Scherbov B (2008) Rethinking age and ageing. Population Bulletin 63(4):1–16

Sauvy A (1948) Social and economic consequences of the ageing of western european populations. Population Studies 2(1):115–124, DOI 10.1080/00324728.1948.10416342

Sündbarg G (1900) Sur la répartition par âge et sur les taux de la mortalité. Bulletin de l'Institut international de statistique 12(1):89–98

Toulemon L (2006) Multidimensional exploratory analysis. In: Caselli D, Vallin J, Wunsch G (eds) Demography - Analysis and Synthesis: A Treatise in Population, Elsevier, London, pp 683–684

Uhlenberg P (2009) International Handbook of Population Aging. Springer, Dordrecht, NL, DOI 10.1007/978-1-4020-8356-3

United Nations (2013) World population ageing 2013. ST/ESA/SER.A/34, department of Economic and Social Affairs

Veyret-Verner G (1971) Populations vieillies. Types, variétés des processus et des incidences sur la population adulte. Revue de géographie alpine 59(4):433–456, DOI 10.3406/rga.1971.1446

Weil DN (1997) The economics of population aging. In: Rosenzweig MR, Stark O (eds) Handbook of Population and Family Economics, Elsevier, New York, pp 967–1014

World Population Prospects (2012) World population prospects: The 2012 revision. http://esa.un.org/wpp/, department of Economic and Social Affairs, United Nations

TCA/HB Compared to CBC/HB for Predicting Choices Among Multi-Attributed Products

Daniel Baier, Marcin Pełka, Aneta Rybicka, and Stefanie Schreiber

Abstract For some years, choice-based conjoint analysis (CBC) has demonstrated its superiority over other preference measurement alternatives. So, e.g., in a recent study on German and Polish cola consumers, the superiority of CBC over traditional conjoint analysis (TCA) was striking. As one reason for this superiority, the usage of hierarchical Bayes for CBC parameter estimation was mentioned (CBC/HB). This paper clarifies whether this really makes the difference: Hierarchical Bayes is also used for TCA parameter estimation (TCA/HB). The application to the above mentioned data shows, that this improves the predictive validity compared to TCA but is still inferior to CBC/HB in "high data quality cases". However, in "low data quality cases" TCA/HB is superior to CBC/HB.

Daniel Baier and Stefanie Schreiber
Brandenburg University of Technology Cottbus-Senftenberg, Chair of Marketing and Innovation Management, Erich-Weinert-Straße 1, 03046 Cottbus, Germany,
✉ daniel.baier@tu-cottbus.de, stefanie.schreiber@tu-cottbus.de

Marcin Pełka and Aneta Rybicka
Wroclaw University of Economics, Department of Econometrics and Computer Science, Nowowiejska 3, 58-500 Jelenia Góra, Poland,
✉ marcin.pelka@ue.wroc.pl, aneta.rybicka@ue.wroc.pl

ARCHIVES OF DATA SCIENCE, SERIES A
KIT SCIENTIFIC PUBLISHING
Vol. 1, No. 1, S. 77–87, 2016

DOI 10.5445/KSP/1000058747/05
ISSN 2363-9881

1 Introduction

In marketing and market research, the application of preference measurement methods for modeling choices among multi-attributed products has a long history. Maybe the best known family of methods is conjoint analysis (CA). CA started its success in the 1960s and 1970s as an approach that allows to estimate part worths for attribute-levels from rankings or ratings of attribute-level-combinations using regression-like procedures (see, e.g., Green and Rao, 1971). Green et al (2001) used in their overview on CA methods the term traditional CA (TCA) for approaches that rely on ranking and/or rating data. It should be mentioned that in TCA the usage of MONANOVA or ANOVA for parameter estimation leads in many cases to rather similar results (see, e.g., Green and Srinivasan, 1978). This is mostly ascribed to the misfit between few observations and many parameters at the desired individual modeling level. Also, in TCA, despite many methodological improvements over the years (e.g. adaptive conjoint analysis), the basic five application steps remained the same (see, e.g., Green et al, 2001):

1. Determine the attributes and levels that influence the customer's choice decisions.
2. Design a set of fictional attribute-level-combinations (stimuli) for data collection.
3. Collect preferential evaluations of these stimuli from a sample of customers.
4. Derive part worths for the attribute-levels using regression-like procedures.
5. Predict the choices of each customer in an assumed market scenario and aggregate them to market shares or sales volumes.

However, today, not TCA but choice-based CA (CBC) (see, e.g., Louviere and Woodworth, 1983; Sawtooth Software, 2013a) is most frequently applied (see, e.g., Selka and Baier, 2014; Selka et al, 2014, for recent overviews on commercial applications). With CBC, in step 3, instead of rating or ranking attribute-level-combinations, the respondents are repeatedly confronted with sets of attribute-level-combinations (so-called choice sets) and asked to select the most preferred ones. Then, in step 4, a multinomial logit model is used for estimation. However, even more severe as TCA, CBC suffers from the misfit between few observations and many parameters at the individual modeling level. As a consequence only pooled models could be estimated (assuming that groups of customers have identical part worths). Here, hierarchical Bayes

(HB) methods for CBC part worth estimation (see, e.g., Allenby and Lenk, 1994; Sawtooth Software, 2009) provided the solution: Observations are shared across respondents during estimation. So, it is possible to estimate individual part worths from few observations per respondent.

Comparison studies (see, e.g., Elrod et al, 1992; Oliphant et al, 1992; Vriens et al, 1998; Moore et al, 1998; Moore, 2004; Karniouchina et al, 2009; Baier et al, 2015) have shown that CBC outperforms TCA in many cases, especially when CBC/HB is used for parameter estimation. However, since also HB methods exist to estimate TCA model parameters (see, e.g., Lenk et al, 1996; Baier and Polasek, 2003; Sawtooth Software, 2013b; Baier, 2014), the question remains unanswered whether CBC/HB is also superior to TCA/HB. This paper tries to close this gap. In Sect. 2 TCA/HB and in Sect. 3 CBC/HB are shortly described. Then, in Sect. 4, the data from Baier et al (2015) are used to compare TCA/HB and CBC/HB. Section 5 closes with a short conclusion and outlook.

2 Hierarchical Bayes Traditional Conjoint Analysis (TCA/HB)

Let $\mathbf{y}_1,\ldots,\mathbf{y}_n \in \mathbb{R}^m$ describe observed preferential evaluations from n respondents ($i = 1,...,n$) w.r.t. to m stimuli ($j = 1,...,m$). y_{ij} denotes the observed preference value of respondent i w.r.t. stimulus j. $\mathbf{X} \in \mathbb{R}^{m\times p}$ denotes the characterization of the m stimuli by p variables. In case of nominal or ordinal attribute-levels a dummy- or an effect-coding is used. The observed evaluations are assumed to come from the following (lower hierarchical) model:

$$\mathbf{y}_i = \mathbf{X}\boldsymbol{\beta}_i + \boldsymbol{\varepsilon}_i, \quad \text{for } i = 1,\ldots,n \quad \text{with } \boldsymbol{\varepsilon}_i \sim N(\mathbf{0},\sigma^2\mathbf{I}), \tag{1}$$

with $\mathbf{I}$ as the identity matrix, σ^2 as an error variance parameter.

The individual part worths $\boldsymbol{\beta}_1,\ldots,\boldsymbol{\beta}_n$ are assumed (higher hierarchical model) to come from a multivariate normal distribution with mean part worth vector $\boldsymbol{\mu} \in \mathbb{R}^p$ and a (positive definite) covariance matrix $\mathbf{H} \in \mathbb{R}^{p\times p}$:

$$\boldsymbol{\beta}_i \sim N(\boldsymbol{\mu},\mathbf{H}) \quad i = 1,\ldots,n. \tag{2}$$

For estimating the model parameters $(\boldsymbol{\mu},\mathbf{H},\boldsymbol{\beta}_1,\ldots,\boldsymbol{\beta}_n,\sigma^2)$, Bayesian procedures provide a mathematically tractable way that combines distributional information about the model parameters with the likelihood of the observed data. The result of this combination, the empirical posterior distribution of the model parameters, is generated by (Gibbs) sampling a sequence of draws from the conditional distributions of the model parameters (see, e.g., Lenk et al, 1996; Baier and Polasek, 2003; Sawtooth Software, 2013b; Baier, 2014, for details) using iteratively the following four steps (starting, e.g., with random values as estimates for the model parameters):

1. Use present estimates of $\boldsymbol{\beta}_1,\ldots,\boldsymbol{\beta}_n$ and $\mathbf{H}$ to generate a new estimate of $\boldsymbol{\mu}$. $\boldsymbol{\mu}$ is assumed to be distributed normally with mean equal to the average of the $\boldsymbol{\beta}_1,\ldots,\boldsymbol{\beta}_n$ and covariance matrix equal to $\mathbf{H}$ divided by the number of respondents. Randomly draw a new estimate of $\boldsymbol{\mu}$ from this distribution.
2. Use present estimates of $\boldsymbol{\beta}_1,\ldots,\boldsymbol{\beta}_n$ and $\boldsymbol{\mu}$ to draw a new estimate of $\mathbf{H}$ from an inverse Wishart distribution in the following way:
   - Calculate $\mathbf{G} = p\mathbf{I} + \sum_{i=1}^{n}(\boldsymbol{\mu}-\boldsymbol{\beta}_i)'(\boldsymbol{\mu}-\boldsymbol{\beta}_i)$.
   - Apply a Cholesky decomposition to $\mathbf{G}^{-1}$ s.t. $\mathbf{G}^{-1} = \mathbf{F}\mathbf{F}'$.
   - Draw $n+p$ vectors $\mathbf{u}_i$ from $N(\mathbf{0},\mathbf{I})$, calculate $\mathbf{S} = \sum_{i=1}^{n+p}(\mathbf{F}\mathbf{u}_i)(\mathbf{F}\mathbf{u}_i)'$.
   - Set $\mathbf{H} = \mathbf{S}^{-1}$.
3. Use present estimates of $\boldsymbol{\mu}$, $\mathbf{H}$, and σ^2 to draw new estimates of $\boldsymbol{\beta}_1,\ldots,\boldsymbol{\beta}_n$ from the following conditional distributions ($i = 1,\ldots,n$):

$$\boldsymbol{\beta}_i \sim N(\boldsymbol{\mu}_i,\mathbf{G}) \text{ with } \mathbf{G} = (\mathbf{H}^{-1}+\sigma^{-2}\mathbf{X}'\mathbf{X})^{-1},\ \ \boldsymbol{\mu}_i = \mathbf{G}(\mathbf{H}^{-1}\boldsymbol{\mu}+\sigma^{-2}\mathbf{X}'\mathbf{y}_i).$$

4. Use present estimates of $\boldsymbol{\mu}$, $\mathbf{H}$, and $\boldsymbol{\beta}_1,\ldots,\boldsymbol{\beta}_n$ to generate a new estimate of σ^2 by a similar – but scalar – approach as in step 2.

The final estimates of the model parameters are obtained by averaging the repeated draws from the above four steps. Here often the draws from the first – so-called burn-in – iterations are omitted.

3 Hierarchical Bayes Choice-Based Conjoint Analysis (CBC/HB)

CBC differs from TCA insofar that respondents are repeatedly confronted with (choice) sets of attribute-level-combinations (stimuli) and asked to select the most preferred one. So, at the (lower hierarchical) level, the choice of stimulus j' out of J alternatives ($j = 1,\ldots,J$) has to be modeled. As usual in multinomial logit models

$$p_{ij'} = \frac{exp(\mathbf{x}_{j'}\boldsymbol{\beta}_i)}{\sum_{j=1} exp(\mathbf{x}_j\boldsymbol{\beta}_i)} \tag{3}$$

is the probability that the respondent i selects stimulus j' (assuming an independently, identically type I extreme distributed additional error in the utilities, see, e.g., Louviere and Woodworth, 1983; Sawtooth Software, 2013a). $\mathbf{x}_j$ denotes the characterization of the alternative j in this choice task ($j = 1,...,J$). As with TCA/HB, the individual part worths $\boldsymbol{\beta}_1,\ldots,\boldsymbol{\beta}_n$ are assumed (higher hierarchical model) to come from a multivariate normal distribution with mean part worth vector $\boldsymbol{\mu} \in \mathbb{R}^p$ and a (positive definite) covariance matrix $\mathbf{H} \in \mathbb{R}^{p\times p}$:

$$\boldsymbol{\beta}_i \sim N(\boldsymbol{\mu},\mathbf{H}) \quad i = 1,\ldots,n. \tag{4}$$

Again, for estimating the model parameters $(\boldsymbol{\mu},\mathbf{H},\boldsymbol{\beta}_1,\ldots,\boldsymbol{\beta}_n)$, Bayesian procedures are used. The steps are similar as above, only the draws w.r.t. the $\boldsymbol{\beta}_1,\ldots,\boldsymbol{\beta}_n$) differ. Here, a Metropolis Hastings algorithm has to be used. For details see Allenby and Lenk (1994) and Sawtooth Software (2009).

4 Empirical studies: Experiments and Results

For testing whether CBC/HB is superior to TCA/HB the data from Baier et al (2015) are used. The multi-attributed product under investigation was – as already mentioned – cola to be bought in the supermarket with the attributes brand (with levels Coca Cola, Pepsi Cola, other brand), flavor (Cola, Cola with orange, Cola with lemon, Cola with cherry), calorie content (normal, light, zero), caffeine content (caffeinated, caffeine-free), price (0.59 €/l, 0.69 €/l, 0.79 €/l, 0.89 €/l), and bottle size (0.5 l, 1 l, 1.5 l, 2 l). The usage of the unit price per volume (€/l) is somewhat problematic since the respondents are used to buy colas at absolute prices (€), but they were explicetely informed about this difference from the usual buying situation.

The data collection took place at two universities near the German-Polish border. The first experiment in each country was an offline-experiment with a TCA task. For TCA, 25 stimuli were generated using orthogonal plans as proposed by SPSS Conjoint to the above number of attributes and levels. In Germany 199 respondents participated in the TCA experiment, in Poland 194. The second experiment in each country was an online-experiment with a CBC task. The respondents were confronted with 18 choice sets, each consisting of four attribute-level-combinations plus a no-choice option. The number of stimuli and choice sets is somewhat high in both experiments, but the students received an incentive and accepted the (complicated) tasks. In Germany 169 respondents participated in the CBC experiment, in Poland 225. All experiments (also the online-experiments) were performed in a controlled laboratory situation: Interviewers informed the respondents about their tasks and observed the answering process. As an incentive for participating, in all experiments, the respondents received a voucher for a small bottle of cola in the cafeteria of their university. All experiments closed with the same holdout choice task (eight identical holdout choice sets were presented) to evaluate the predictive validity. All experiments were performed during four weeks in May and June 2013.

As discussed in Baier et al (2015), the data collection in all four experiments was possible without problems. However, in all experiments, data inspection showed that there were cheating respondents when filling out the questionnaires: Some obviously didn't sort the stimuli with great efforts (resulting in "similar" orderings compared to the stimuli numbers), some used some simplifying rules when selecting stimuli in the choice sets (e.g. always selecting the first stimulus in the set). The total number of such directly observable cheaters was small (e.g. about 10 % in the Polish samples, 5 % in the German samples), but this supported the impression that the Polish samples were data "of lower quality" than the German samples.

The TCA data were analyzed in Baier et al (2015) using MONANOVA as implemented in SPSS Conjoint whereas the CBC data were analyzed using Sawtooth Software's CBC/HB software (Sawtooth Software, 2009). The "directly observed" low data quality of the Polish samples is reflected in the model fit: So, e.g., 16 from the 194 respondents in the Polish TCA sample showed a Pearson correlation of 0.7 or lower when comparing the observed and the estimated preference values. In the German TCA sample only three respondents showed such a low model fit. On average, the Polish respondents showed a Pearson correlation of 0.858 whereas the German respondents showed a Pearson correlation of 0.982. Similar differences can be observed between the Polish and German CBC samples. So, we refer to the two German experiments in the following as a "high data quality case" whereas the two Polish experiments are referred to as the "low data quality case". Especially in the "high data quality case", the CBC experiment showed in Baier et al (2015) a clear superiority w.r.t. predictive validity.

Now, for the research question in this paper, we analyzed also the TCA data with hierarchical Bayes procedures. We used Sawtooth Software's Hierarchical Bayes Regression software for this purpose (Sawtooth Software, 2013b) with 50,000 draws as burn-ins and 10,000 draws for calculating the parameter estimates. We used constraints w.r.t. the price levels in order to prevent

Table 1 Averaged standardized part worths and attribute importances for the different samples in Germany and Poland (TCA/HB=Hierarchical Bayes Traditional Conjoint Analysis, CBC/HB=Hierarchical Bayes Choice-Based Conjoint Analysis); sample differences between Germany and Poland were t-tested; *: significant at α=.05, **: at α=.01, ***: at α=.001

		Averaged standardized part worths (std. dev.)							
		Germany				Poland			
Attribute	Level	TCA/HB (n=199)		CBC/HB (n=169)		TCA/HB (n=194)		CBC/HB (n=225)	
Brand	Coca Cola	.181	(.201)	.172	(.127)	.141	(.157)	.144	(.172)
	Pepsi Cola	.072	(.108)	.075	(.073)	.119	(.144)	.121	(.141)
	Other	.091***	(.141)	.015	(.035)	.120*	(.156)	.088	(.164)
Flavor	Cola	.236	(.188)	.241	(.140)	.111	(.137)	.164***	(.162)
	W. orange	.119	(.152)	.094	(.111)	.055	(.078)	.072	(.109)
	W. lemon	.153	(.146)	.132	(.117)	.125	(.130)	.105	(.111)
	W. cherry	.089*	(.124)	.060	(.100)	.150***	(.115)	.101	(.135)
Calorie	Normal	.167	(.156)	.193	(.167)	.086**	(.084)	.064	(.080)
	Light	.064	(.097)	.086*	(.081)	.086***	(.085)	.037	(.056)
	Zero	.042	(.076)	.046	(.096)	.037	(.067)	.048	(.071)
Caffeine	Caffein.	.128***	(.138)	.083	(.092)	.039	(.062)	.035	(.049)
	C.-free	.009	(.036)	.014	(.045)	.034	(.049)	.026	(.060)
Price	.59 €/l	.029	(.072)	.087***	(.095)	.097	(.071)	.130*	(.148)
	.69 €/l	.022	(.049)	.060***	(.065)	.031	(.037)	.100***	(.074)
	.79 €/l	.011	(.029)	.038***	(.044)	.002	(.001)	.086***	(.079)
	.89 €/l	.000	(.000)	.013***	(.025)	.000	(.000)	.044***	(.094)
Bottle size	0.5 l	.085***	(.093)	.027	(.049)	.049	(.059)	.056	(.090)
	1 l	.079*	(.085)	.061	(.052)	.106**	(.090)	.059	(.075)
	1.5 l	.054	(.064)	.049	(.050)	.112**	(.094)	.071	(.075)
	2 l	.033	(.051)	.048**	(.056)	.093	(.090)	.075	(.108)

	Averaged attribute importances (std. dev.)							
	Germany				Poland			
Attribute	TCA/HB		CBC/HB		TCA/HB		CBC/HB	
Brand	.231*	(.197)	.187	(.116)	.254	(.160)	.256	(.192)
Flavor	.292	(.180)	.279	(.133)	.256	(.125)	.255	(.170)
Calorie	.189	(.150)	.241***	(.141)	.136***	(.083)	.105	(.086)
Caffeine	.138***	(.134)	.096	(.091)	.073*	(.061)	.061	(.065)
Price	.029	(.072)	.103***	(.090)	.097	(.071)	.186***	(.140)
Bottle size	.122***	(.090)	.094	(.061)	.184***	(.080)	.137	(.121)

degeneration. The internal validity of TCA/HB was 0.767 (averaged R^2 across respondents) for the German sample and 0.418 for the Polish sample. Please note, as in Baier et al (2015) for TCA and CBC/HB, the “lower quality" of the Polish data. The internal validity of CBC/HB (see Baier et al, 2015) was 0.647 (averaged root likelihood value across respondents) for the German and 0.598 for the Polish sample. However, the respondents with a low model fit were not removed from the further analyses since we explicetely wanted to demonstrate the ability of the different procedures to deal with the “low data quality" problem.

Afterwards, for comparison reasons, the estimated part worths at the respondent level were standardized in the usual way so that – for each respondent in each experiment – the maximum possible value for a stimulus is 1 and the minimum 0. Also the individual attribute importances in each experiment were calculated via the difference of the highest and the lowest part worth for levels of the corresponding attribute. Table 1 gives the averaged part worths for all four experiments (TCA/HB, CBC/HB in Germany and Poland) and also averaged importances. From Table 1 one can easily see, that – more or less – the results across nationality (also: “data quality", see above) and methods are similar: “Flavor" is – on average – the most important attribute when selecting colas, followed by “brand" and “calorie". The importance of “price" differs between German and Polish consumers but also between TCA/HB and CBC/HB. So, e.g., with CBC/HB, the importance of price is much higher than with TCA/HB. This can be partly ascribed to the well-known fact that “simple" or “quantifiable" attributes are more looked at in the choice than in the ranking setting (see, e.g., Karniouchina et al, 2009; Baier et al, 2015), but also to the constraining of the price parameters during TCA/HB estimation.

The most important comparison deals with the predictive validity. Here the responses to the holdout choice tasks (identical in all experiments) have to be compared with model predictions (assuming that the respondent selects the stimulus with highest predicted sum of part worths in each holdout choice set). There are two possibilities to calculate them: One could control all holdout choices (including the selections of the no-choice option, in total n=9,880 selections) or one could control only the holdout choices where a holdout stimulus was selected (excluding the selection of the no-choice option, in total n=7,803 selections). The fair comparison would be the second one (see, e.g., Karniouchina et al, 2009; Baier et al, 2015), since TCA resp. TCA/HB do not collect data to predict the no-choice option, only CBC resp. CBC/HB are able to give such predictions.

Table 2 First Choice Hit Rates (FCHR). (TCA/HB=Hierarchical Bayes Traditional Conjoint Analysis, CBC/HB=Hierarchical Bayes Choice-Based Conjoint Analysis); "With NC" stands for all choices of respondents in the holdout tasks including the no-choice selections, for "Without NC" the no-choice selections are excluded; a binomial test was applied to compare the results: FCHR of one method was assumed and checked whether FCHR of the other is higher; *: significant at α=.05, **: at α=.01, ***: at α=.001

	First Choice Hit Rates			
	Germany		Poland	
Holdout choices considered	TCA/HB	CBC/HB	TCA/HB	CBC/HB
With NC	.456	.710***	.339	.335
Without NC	.648	.726*	.378	.323

So, for the second one, in Germany, CBC/HB outperforms TCA/HB with First Choice Hit Rate (FCHR, the percentage of correctly predicted selections in the holdout choice sets) values of 0.726 (for CBC/HB) and 0.648 (for TCA/HB). The TCA/HB value is better than the value 0.626 for TCA in Baier et al (2015), but still worse than CBC/HB. See Table 2. In the Polish experiments, TCA/HB (FCHR=0.378) outperforms CBC/HB (FCHR=0.323). See Table 2. However, since the FCHR values for the Polish experiments are very low, one should nevertheless conclude that CBC is superior with respect to prediction, especially in "high data quality cases".

5 Conclusions and outlook

The analyses have shown, that – especially in "high data quality cases" – for choice predictions, CBC/HB outperforms TCA and also – as a result of this paper – TCA/HB. However, especially in markets with "low data quality", the TCA/HB approach competes well, especially when the no-choice options are neglected. Of course, the comparisons between traditional and choice-based methods have not come to an end, one needs more such comparisons.

However, already from the comparison in this paper, one can draw ideas for methodological improvements: The main improvement with respect to predictive accuracy but also with respect to model fit comes from the motivation of the respondents to answer the questionnaires carefully. The "low data quality" problem in the Polish samples can only partly be compensated by the better HB estimation procedure. So, it seems that improvements that focus on better data collection and respondent motivation procedures are of higher value for marketing research practice than better parameter estimation procedures.

References

Allenby GM, Lenk PJ (1994) Modeling household purchase behavior with logistic regression. Journal of the American Statistical Association 89(428):1218–1231

Baier D (2014) Bayesian Methods for Conjoint Analysis-Based Predictions: Do We Still Need Latent Classes?, Springer International Publishing, Cham, pp 103–113. DOI 10.1007/978-3-319-01264-3_9

Baier D, Polasek W (2003) Market simulation using Bayesian procedures in conjoint analysis. In: Schwaiger M, Opitz O (eds) Exploratory Data Analysis in Empirical Research, Studies in Classification, Data Analysis, and Knowledge Organization, Springer, Berlin, pp 413–421, DOI 10.1007/978-3-642-55721-7_42

Baier D, Pełka M, Rybicka A, Schreiber S (2015) Ratings-/Rankings-Based Versus Choice-Based Conjoint Analysis for Predicting Choices, Springer, Berlin, pp 205–216. DOI 10.1007/978-3-662-44983-7_18

Elrod T, Louviere JJ, Davey K (1992) An empirical comparison of ratings-based and choice-based conjoint models. Journal of Marketing Research 29(3):368–377

Green PE, Rao VR (1971) Conjoint measurement for quantifying judgmental data. Journal of Marketing Research 8(3):355–363

Green PE, Srinivasan V (1978) Conjoint analysis in consumer research: Issues and outlook. Journal of Consumer Research 5(2):103–123

Green PE, Krieger AM, Wind Y (2001) Thirty years of conjoint analysis: Reflections and prospects. Interfaces 31(3b):56–73

Karniouchina E, Moore WL, Van der Rhee B, Verma R (2009) Issues in the use of ratings-based versus choice-based conjoint analysis in operations management research. European Journal of Operational Research 197(1):340–348

Lenk PJ, DeSarbo WS, Green PE, Young MR (1996) Hierarchical Bayes conjoint analysis: Recovery of partworth heterogeneity from reduced experimental designs. Marketing Science 15(2):173–191

Louviere JJ, Woodworth G (1983) Design and analysis of simulated consumer choice or allocation experiments: An approach based on aggregate data. Journal of Marketing Research 20(4):350–367

Moore WL (2004) A cross-validity comparison of ratings-based and choice-based conjoint analysis models. International Journal of Research in Marketing 21(3):299–312, DOI 10.1016/j.ijresmar.2004.01.002

Moore WL, Gray-Lee J, Louviere JJ (1998) A cross-validity comparison of conjoint analysis and choice models at different levels of aggregation. Marketing Letters 9(2):195–207, DOI 10.1023/A:1007913100332

Oliphant K, Eagle T, Louviere J, Anderson D (1992) Cross-task comparison of ratings-based and choice-based conjoint. In: Metegrano M (ed) Sawtooth Software Conference Proceedings, Sawtooth Software Inc., Ketchum, Idaho, pp 383–404

Sawtooth Software (2009) The CBC/HB System for Hierarchical Bayes Estimation Version 5.0 Technical Paper. Sawtooth Software Inc., Sequim, Washington, URL `http://www.sawtoothsoftware.com/support/technical-papers/hierarchical-bayes-estimation/cbc-hb-technical-paper-2009`

Sawtooth Software (2013a) The CBC System for Choice-Based Conjoint Analysis Version 8. Sawtooth Software Inc., Orem, Utah

Sawtooth Software (2013b) HB-Reg v4 for Hierarchical Bayes Regression. Sawtooth Software Inc., Orem, Utah

Selka S, Baier D (2014) Kommerzielle Anwendung auswahlbasierter Verfahren der Conjointanalyse: Eine empirische Untersuchung zur Validitätsentwicklung. Marketing ZFP 36(1):54–64

Selka S, Baier D, Kurz P (2014) The validity of conjoint analysis: An investigation of commercial studies over time. In: Spiliopoulou M, Schmidt-Thieme L, Janning R (eds) Data Analysis, Machine Learning and Knowledge Discovery, Springer International Publishing, Cham, pp 227–234, DOI 10.1007/978-3-319-01595-8_25

Vriens M, Oppewal H, Wedel M (1998) Rating-based versus choice-based latent class conjoint models: An empirical comparison. Journal of the Market Research Society 40(3):237–248

Maximum Difference Scaling Method in the `MaxDiff` R Package

Tomasz Bartłomowicz and Andrzej Bąk

Abstract In microeconomics, measurement of consumer preferences is one of the most important elements of marketing research. Accurate measurement of preferences allows to gain an understanding of likes and dislikes of consumers. Using some statistical methods (like e.g. conjoint analysis and discrete choice models) it is possible to quantify preferences and answer the questions: What product will a consumer choose? What attribute of the product is most important? Consumer choice models attempt to answer these questions. This article describes the R package `MaxDiff` for the Maximum Difference Scaling method to assess consumer preferences from consumer choice experiments. Because practical applications of this method depend on the availability of computer software, this paper describes an implementation of the Maximum Difference Scaling method in the R package `MaxDiff`. Functions of the `MaxDiff` R package can be used for the measurement of consumer preferences. `MaxDiff` supports the design of the experiment (e.g. to build a list of features), encode the alternatives, estimate the models, etc. Some functions of

Tomasz Bartłomowicz
Wrocław University of Economics, Department of Econometrics and Computer Science, Nowowiejska 3, 58-500 Jelenia Góra, Poland,
✉ `tomasz.bartlomowicz@ue.wroc.pl`

Andrzej Bąk
Wrocław University of Economics, Department of Econometrics and Computer Science, Nowowiejska 3, 58-500 Jelenia Góra, Poland
✉ `andrzej.bak@ue.wroc.pl`

ARCHIVES OF DATA SCIENCE, SERIES A
KIT SCIENTIFIC PUBLISHING
Vol. 1, No. 1, S. 89–101, 2016

DOI 10.5445/KSP/1000058747/06
ISSN 2363-9881

the `MaxDiff` R package are presented with examples of applications in the empirical analysis of consumer preferences.

1 Introduction

Maximum Difference Scaling (MaxDiff) is a relatively new approach for measuring the importance of preferences for multiple items like product features, job-related benefits, advertising claims, product packaging, etc. Although Maximum Difference Scaling has much in common with conjoint analysis and discrete choice methods, the method is easier to use for researchers, respondents and clients. We can say, that the MaxDiff method combines the best features of traditional conjoint analysis and discrete choice methods. That is why Maximum Difference Scaling is also known as Best-Worst Scaling or Best-Worst Conjoint (Louviere, 1991). Comparison of the most popular preferenc measurement methods is presented in Table 1.

Maximum Difference Scaling (originally Best-Worst Scaling) is a preference measurement method developed by Louviere and and other researchers (see Louviere and Woodworth, 1983; Louviere, 1991; Finn and Louviere, 1992; Marley and Louviere, 2005).

With the MaxDiff method, respondents are shown subsets of the possible items in the experiment and are asked to indicate (among these subsets) the most and the least preferred (best and worst) items. Respondents typically evaluate a dozen sets where each set contains a different subset of items. The combinations of items are designed very carefully. Each item is shown with an equal number of pairs of items an equal number of times. Each respondent typically sees each item two or even more times across the MaxDiff sets. Compared to the rankings which is usually limited to a small number if items and to the scaled ratings, MaxDiff choosing is usually selective enough. Let us consider a set in which a respondent evaluates five items: A, B, C, D and E. If the respondent says that A is the best and E is the worst, these two responses inform us on seven of ten possible implied paired comparisons: A>B, A>C, A>D, A>E, B>E, C>E, D>E. In the opinion of some authors, humans are much better at judging items at extremes than in discriminating among items of middling importance of preference. Maximum Difference Scaling experiments focus on estimating preference or importance scores for typically about 15 to 30 attributes (Sawtooth Software, 2013).

Table 1 Comparison of the most popular methods of preferences measurement

Specification	Traditional conjoint analysis	Discrete choice method	Maximum Difference Scaling
Number of variables (attributes)	Max 6 attributes	Max 9 attributes	15-30 attributes
Method of profiling	Full factorial design, fractional factorial design	Blocking factorial design	Subsets of items
Method of data collection	Ranking (rating) of all profiles	Choosing of the most preferred item or any one	Choosing of the most and least preferred (best and worst) items
Model	Multiple regression model	Multinomial, conditional, mixed logit model	Multinomial logit model
Estimation method	OLS regression	Maximum likelihood method, Expectation-Maximization (EM) algorithm	Maximum likelihood method
Commercial software	SPSS, STATISTICA, Sawtooth Software	SAS/STAT, STATISTICA, S-PLUS	Sawtooth Software
Free software (GNU GPL license)	conjoint R package	DiscreteChoice R package, mlogit R package	MaxDiff R package

In MaxDiff models estimation of the utility function is typically performed using multinomial discrete choice models, in particular multinomial logit models. Several algorithms could be used in this estimation process, including maximum likelihood, neural networks and the hierarchical Bayes method. In the `MaxDiff` R package a multinomial logit model with maximum likelihood estimation method is used. Additional information about the MaxDiff method can be found in (Cohen, 2003; Louviere, 1991; Sawtooth Software, 2013).

2 The `MaxDiff` R package functions

The `MaxDiff` package is an implementation of the Maximum Difference Scaling method for R (Bartłomowicz and Bąk, 2013). The package is available under the GNU General Public License with free access to source code. The current version of the `MaxDiff` package is 1.12. It is possible to download the package from the CRAN packages repository[1] and the home WWW page of the Department of Econometrics and Computer Science of the Wrocław University of Economics[2]. To use the package it is necessary to install the base R computer program (R Development Core Team, 2013) and two other packages: `mlogit` (Croissant, 2012) for estimation of the logit models and `AlgDesign` (Wheeler, 2004) for generating fractional factorial designs.

The current version of the `MaxDiff` package (v. 1.12) has thirteen functions. All of them (with their arguments and short description) are presented in Table 2 (in order of the MaxDiff procedure).

The first two functions are used to make fractional factorial design with the suggested number of profiles and alternatives in each block of profiles using vector (or matrix) of alternatives' names. The function `mdBinaryDesign()` returns a binary fractional factorial design while the function `mdAggregate-Design()` returns an aggregate fractional factorial design. If we want to make a design with alternatives' names, it is necessary to use next the `mdDesign-Names()` function which replaces the binary or aggregate fractional factorial design in the design with the alternatives' names.

The next two functions: `mdAggregateToBinaryDesign()` and `mdBinaryToAggregateDesign()` convert binary designs to aggregate designs or aggregate designs to binary ones. These functions are complements of each other, because for some of the functions (`mdRankData()`, `mdLogitData()`, `mdLogitIndividualCounts()`, `mdLogit-Individual-Ranks()`, `mdMeanRanks()`, `mdLogitModel()`, `mdLogitRanks()` and `mdMeanIndividualCounts()`) it is necessary to convert an aggregate design to a binary design.

In the group of data set functions there are two functions, namely the function `mdRankData()` which converts a basic data set into a rank data set and the function `mdLogitData()` which converts a rank data set into a logit data set. In the first case the basic data set from questionnaires is converted into a special data set for almost all functions of the `MaxDiff` package except the function `mdLogitModel()`. For this last function, it is necessary to use the `mdLogitData()` function which converts rank data into a special data set for the logit model which is estimated with the `mlogit` R package.

In the next group of functions we find the function `mdMeanIndividual-Counts()` and the function `mdMeanRanks()`. The first of them computes

[1] http://cran.r-project.org/web/packages/MaxDiff

[2] http://keii.ue.wroc.pl/MaxDiff/

Table 2 Functions of `MaxDiff` R package with required arguments

Function header and description
`mdBinaryDesign(profiles.number, alternatives.per.profile.number, alternatives.names)` function makes binary fractional factorial design with suggested number of profiles and alternatives in each profile using vector (or matrix) of alternatives' names
`mdAggregateDesign(profiles.number, alternatives.per.profile.number, alternatives.names)` function makes aggregate fractional factorial design with suggested number of profiles and alternatives in each profile using vector (or matrix) of alternatives' names
`mdDesignNames(binary.or.aggregate.design, alternatives.names)` function replaces binary or aggregate fractional factorial design in design with alternatives' names
`mdAggregateToBinaryDesign(aggregate.design, alternatives.names)` function converts aggregate design to binary design with alternatives' names
`mdBinaryToAggregateDesign(binary.design)` function converts binary design to aggregate design
`mdRankData(basic.data, binary.design)` function converts basic data set into rank data set for functions: *mdIndividualCounts(), mdLogitData(), mdLogitRanks(), mdLogitIndividualCounts(), mdLogitIndividualRanks(), mdMeanRanks()*
`mdLogitData(rank.data, binary.design, alternatives.names)` function converts rank data set into logit data set for *mdLogitModel()* function
`mdMeanIndividualCounts(rank.data, binary.design)` function computes the individual-level counts for each respondents
`mdMeanRanks(rank.data, binary.design)` function computes the overall counts for the whole sample using the arithmetic mean
`mdLogitModel(logit.data, binary.design, alternatives.names)` function estimates an aggregate logit model
`mdLogitRanks(rank.data, binary.design, alternatives.names)` function computes the overall counts and ranks for the whole sample using the logit model
`mdLogitIndividualCounts(rank.data, binary.design, alternatives.names)` function computes the individual-level counts for each respondent using the logit model
`mdLogitIndividualRanks(rank.data, binary.design, alternatives.names)` function computes the individual-level ranks for each respondent using the logit model

Arguments of functions	
`profiles.number`	Number of profiles in every block
`alternatives.per.profile.number`	Number of alternatives in every block of profiles
`alternatives.names`	Vector (or matrix) with alternatives' names
`binary.or.aggregate.design`	Binary or aggregate fractional factorial design
`aggregate.design`	Aggregate fractional factorial design
`binary.design`	Binary fractional factorial design
`basic.data`	Data set from questionnaires
`rank.data`	Data set with ranks
`logit.data`	Data set for logit model

the individual level counts for each respondent, while the second one computes the overall counts for the whole sample using the arithmetic mean.

The last group of functions is linked to the logit model. In this group there are four functions. The first function – `mdLogitModel()` estimates an aggregate logit model. The second one – `mdLogitRanks()` computes the overall counts and ranks for the whole sample using the logit model. The third one, the function `mdLogitIndividualCounts()`, computes the individual level counts for each respondent using the logit model and the fourth function `mdLogitIndividualRanks()` computes the individual-level ranks for each respondent using the logit model.

The detailed description and more examples of the use of all functions are available in the documentation of the `MaxDiff` R package (Bartłomowicz and Bąk, 2013).

3 The `MaxDiff` R package application

In the application example of the `MaxDiff` R package the identification and analysis of the preferences of respondents using some forms of job benefits is proposed. The main aim was to determine the most and least important features of the following job benefits: phone (mobile), laptop, company car, voucher, house subsidy and food subsidy. The data set of job benefits choice data allows to illustrate the use of the `MaxDiff` R package.

In the example, the job benefits experiment has 6 choice options. This means that in the outcome it was necessary to build a fractional factorial design with at least 5 profiles of job benefits. In the `MaxDiff` R package it is possible to generate a binary design (for the rest of calculations) and an aggregate design (used in the questionnaires) with profiles as a fractional factorial design. In the following example 5 profiles with 3 (from 6) attributes in each profile were generated[3]:

[3] The same fractional factorial design is saved as matrix `X` in the `Job_benefits` sample data set for `MaxDiff` R package.

```
> library(MaxDiff)
> Z=c("Phone","Laptop","Company_car","Voucher","House_subsidy",
        "Food_subsidy")
> X=mdBinaryDesign(5, 3, Z)
> print(X)
                   Profile1 Profile2 Profile3 Profile4 Profile5
Phone              1        0        0        1        1
Laptop             1        1        0        0        0
Company_car        0        1        0        1        0
Voucher            0        1        1        0        1
House_subsidy      0        0        1        1        1
Food_subsidy       1        0        1        0        0
```

The binary design can also be converted into an aggregate design with the help of the function mdBinaryToAggregateDesign() function[4]:

```
> X.aggregate=mdBinaryToAggregateDesign(X)
> print(X.aggregate)

  Profile1 Profile2 Profile3 Profile4 Profile5
1        1        2        4        1        1
2        2        3        5        3        4
3        6        4        6        5        5
```

Besides that, it is possible to create an aggregate design immediately with the function mdAggregateDesign(). To see the design in the form of a questionnaire we should replace the numbers with attributes' names using the function mdDesignNames(). It does not matter if we use the binary or the aggregate design as the primary parameter:

```
> survey.design=mdDesignNames(X.aggregate, Z)
> print(survey.design)

  Profile1     Profile2    Profile3      Profile4      Profile5
1 Phone        Laptop      Voucher       Phone         Phone
2 Laptop       Company_car House_subsidy Company_car   Voucher
3 Food_subsidy Voucher     Food_subsidy  House_subsidy House_subsidy
```

To present and check the MaxDiff R package there should be used some data. In the example we use an artificial data set for 10 respondents. The data set contains the choice of the best and worst attribute in each profile for each respondent (three attributes) and is shown below. For example, for the first respondent in the first profile the best attribute is phone (coded as 1) and the worst attribute is food subsidy (coded as 6).

[4] It is also possible to convert the prepared aggregate design into a binary design with mdAggregate-ToBinaryDesign() function.

```
> library(MaxDiff)
> data(Job_benefits)
> print(Y)
   Id Profile Best Worst
1  1       1    1     6
2  1       2    3     2
3  1       3    6     4
4  1       4    3     5
5  1       5    1     4
6  2       1    1     6
7  2       2    2     4
8  2       3    5     6
9  2       4    1     5
10 2       5    1     5
...
```

To calculate the next functions it was necessary to convert the data matrix `Y` shown above into a rank data set:

```
> rank.data=mdRankData(basic.data=Y, binary.design=X)
> print(rank.data)
         Phone Laptop Company_car Voucher House_subsidy Food_subsidy
Profile1     1      0          NA      NA            NA           -1
Profile2    NA     -1           1       0            NA           NA
Profile3    NA     NA          NA      -1             0            1
Profile4     0     NA           1      NA            -1           NA
Profile5     1     NA          NA      -1             0           NA
Profile1     1      0          NA      NA            NA           -1
...
```

A rank data set represents each alternative as a variable, with missing value codes (`NA`) used when alternatives are not shown, a `1` is used to denote an alternative that is chosen as best, `-1` for worst and `0` for alternatives shown but neither best nor worst (not chosen). This structure of data is a very useful way of setting up results in statistical programs. When data is structured in this way the counts can be, for example, computed using sums and arithmetic mean[5]:

```
> mean.ranks=mdMeanRanks(rank.data, binary.design=X)
> mean.ranks
                   Counts Ranks
Phone           0.4000000     1
Laptop          0.3500000     2
Company_car     0.3000000     3
Voucher        -0.2333333     4
House_subsidy  -0.4000000     6
Food_subsidy   -0.3000000     5
```

[5] If we want to compute the individual-level counts for each respondents we should use `mdMeanIndividualCounts()` function.

With the mdMeanRanks() function it is possible to rank the attributes. According to this ranking, the most attractive are the following job-benefits: phone, then laptop and company car. The least attractive are: voucher, food subsidy and house subsidy. But because the design can be unbalanced (some attributes can be shown more times than others) it is rather a very simple way to rank the attributes. That is, why a logit model should rather be used to count and rank the attributes.

First of all, the rank data should be converted into a logit data set[6] (Food_subsidy abbreviated to Food...):

```
> logit.data=mdLogitData(rank.data, binary.design=X,
        alternatives.names=Z)
> print(head(logit.data))
  ID Set Choice Phone Laptop Company_car Voucher House_subsidy Food...
1  1   1      1     1      0           0       0             0       0
2  1   1      0     0      1           0       0             0       0
3  1   1      0     0      0           0       0             0       1
4  1   2      0    -1      0           0       0             0       0
5  1   2      0     0     -1           0       0             0       0
6  1   2      1     0      0           0       0             0      -1
...
```

The data set logit.data allows to use the mlogit R package in the MaxDiff R package to estimate the logit model:

```
> mdLogitModel(logit.data, binary.design=X, alternatives.names=Z)

Call:
mlogit(formula = formula, data = logit.data, alt.levels =
    paste(1:alternatives.per.profile.number),
    shape = "long", method = "nr", print.level = 0)

Frequencies of alternatives:
   1    2    3
0.33 0.30 0.37

nr method
4 iterations, 0h:0m:0s
g'(-H)^-1g = 0.00139
successive function values within tolerance limits

Coefficients :
               Estimate Std. Error t-value  Pr(>|t|)
Laptop         0.167594   0.436420  0.3840 0.7009635
Company_car    0.032706   0.433899  0.0754 0.9399142
Voucher       -1.413223   0.408347 -3.4608 0.0005385 ***
House_subsidy -1.752940   0.402960 -4.3502  1.36e-05 ***
```

6 Based on: http://surveyanalysis.org/wiki/Analyzing_Max-Diff_Using_-Standard_Logit_Models_Using_R.

```
Food_subsidy  -1.528994   0.455613 -3.3559 0.0007911 ***
---
Signif. codes:  0 '***' 0.001 '**' 0.01 '*' 0.05 '.' 0.1 ' ' 1

Log-Likelihood: -90.622
```

The logit model estimates all parameter values relative to the first alternative (attribute phone), where the alternative has a parameter of 0. It means that 2 attributes (laptop and company car) are more attractive than phone, and 3 attributes (voucher, food subsidy and house subsidy) are not so attractive as a phone.

Similar results can be reached with the function `mdLogitRanks()`. This function computes the overall counts and ranks for the whole sample using a logit model:

```
> logit.ranks=mdLogitRanks(rank.data, binary.design=X,
        alternatives.names=Z)
> print(logit.ranks)
             Counts Rank
Phone          15.5    3
Laptop         41.4    1
Company_car    32.3    2
Voucher         1.1    6
House_subsidy   3.4    5
Food_subsidy    6.3    4

> print(logit.ranks[order(logit.ranks[, 2]), ])
             Counts Rank
Laptop         41.4    1
Company_car    32.3    2
Phone          15.5    3
Food_subsidy    6.3    4
House_subsidy   3.4    5
Voucher         1.1    6

> sum(logit.ranks[, 1])
[1] 100
```

Thus, between the attributes presented in the example there is the following MaxDiff relationship: laptop>company car>phone>food subsidy>house subsidy> voucher. This result is achieved for the whole sample (10 respondents). It is also possible to compute the individual counts and ranks for individual respondents using the logit model. The function `mdLogitIndividualRanks()` computes individual ranks for each respondent:

```
> mdLogitIndividualRanks(rank.data, binary.design=X,
        alternatives.names=Z)
      Phone Laptop Company_car Voucher House_subsidy Food_subsidy
 [1,]     2      5           1       6             4            3
 [2,]     1      2           3       5             4            6
 [3,]     2      1           4       3             6            5
 [4,]     1      5           2       4             3            6
 [5,]     6      3           1       5             2            4
 [6,]     2      1           5       3             6            4
 [7,]     2      3           1       6             5            4
 [8,]     4      3           2       5             6            1
 [9,]     2      1           5       3             6            4
[10,]     4      1           2       5             3            6
```

For most respondents, the phone is attractive or very attractive but most attractive is the laptop. Some respondents prefer other job benefits. For example, for 5th respondent, the phone is the least attractive and the company car is the best option. The food subsidy is the most attractive option for the 8th respondent, but only for him, not for the whole sample. That is why, for the whole sample these 3 attributes (food subsidy, house subsidy, and voucher) are the worst.

4 Conclusions

The MaxDiff package presented in this article is a new package for R designed mostly for statisticians, econometricians, economists and students of economics who are interested in the research of stated consumers preferences. As the R environment and many other packages for R, the package is available for free (under the GNU General Public License with free access to source code). Nevertheless, it is as useful as commercial specialized computer software packages.

The MaxDiff R package implements the Maximum Difference Scaling method supporting all steps of the method. It is possible to design the experiment, encode the alternatives, estimate the models, etc. within the same environment – the MaxDiff R package. By building a binary or aggregate fractional factorial design with the suggested number of profiles and alternatives it is also possible to generate a useful questionnaire for respondents in the package. In the authors' opinion, the MaxDiff R package provides an integrated support of the process of a Maximum Difference Scaling experiment for the researchers.

Table 3 R packages for measurement of stated preferences from Department of Econometrics and Computer Science Wrocław University of Economics

Package name	Implemented method	Authors	Download site
`conjoint`	Traditional *conjoint analysis* (Bąk and Bartłomowicz, 2013a)	Andrzej Bąk, Tomasz Bartłomowicz	CRAN: `http://cran.r-project.org/web/packages/conjoint/` Homepage: `http://keii.ue.wroc.pl/conjoint/`
`Discrete-Choice`	Discrete choice method (Bąk and Bartłomowicz, 2013b)	Andrzej Bąk, Tomasz Bartłomowicz	CRAN: `http://cran.r-project.org/web/packages/DiscreteChoice/` Homepage: `http://keii.ue.wroc.pl/DiscreteChoice/`
`MaxDiff`	Maximum Difference Scaling	Tomasz Bartłomowicz, Andrzej Bąk	CRAN: `http://cran.r-project.org/web/packages/MaxDiff/` Homepage: `http://keii.ue.wroc.pl/MaxDiff/`

In the current version the MaxDiff R package contains a mix of own functions and of functions of packages maintained by others to implement the Maximum Difference Scaling method. It means that to use the package it is necessary to install, in addition to the base R computer program, the `mlogit` and `AlgDesign` packages. This makes the MaxDiff R package dependent on the `mlogit` and the `AlgDesign` package. Because it is not the first package for measurement of stated preferences implemented by members of the Department of Econometrics and Computer Science of the Wroclaw University of Economics (see Table 3), the authors have some experience as maintainers of the packages: In the experience of the authors a new software version of the `mlogit` and the `AlgDesign` package often require also a software update in the MaxDiff R package. And because of this, it is desirable that in the future the MaxDiff R package does not depend on any external packages.

References

Bartłomowicz T, Bąk A (2013) Maximum Difference Scaling – package MaxDiff. URL http://keii.ue.wroc.pl/MaxDiff/

Bąk A, Bartłomowicz T (2013a) Conjoint analysis – package conjoint. URL http://cran.r-project.org/web/packages/conjoint

Bąk A, Bartłomowicz T (2013b) Discrete choice methods – package DiscreteChoice. URL http://keii.ue.wroc.pl/DiscreteChoice/

Cohen SH (2003) Maximum difference scaling: Improved measures of importance and preference for segmentation. Tech. rep., Sawtooth Software Conference Proceedings, URL http://www.sawtoothsoftware.com/download/techpap/maxdiff.pdf

Croissant Y (2012) Multinomial logit model – package mlogit. URL http://cran.r-project.org/web/packages/mlogit

Finn A, Louviere JJ (1992) Determining the appropriate response to evidence of public concern: The case of food safety. Journal of Public Policy & Marketing 11(2):12–25

Louviere JJ (1991) Best-worst scaling: A model for the largest difference judgments, Working Paper, University of Alberta

Louviere JJ, Woodworth G (1983) Design and analysis of simulated consumer choice or allocation experiments: An approach based on aggregate data. Journal of Marketing Research 20(4):350–367

Marley A, Louviere J (2005) Some probabilistic models of best, worst, and best–worst choices. Journal of Mathematical Psychology 49(6):464–480, DOI 10.1016/j.jmp.2005.05.003

R Development Core Team (2013) R: A language and environment for statistical computing, R foundation for statistical computing. URL http://cran.r-project.org/

Sawtooth Software (2013) What is MaxDiff? URL http://www.sawtoothsoftware.com/products/maxdiff-software/93-support/sales-support/238-maxdiff-method

Wheeler RE (2004) eval.design. AlgDesign. The R project for statistical computing, URL http://www.r-project.org/

Various Approaches to Measuring Effectiveness of Tertiary Education

Józef Dziechciarz, Marta Dziechciarz-Duda, Anna Król and Marta Targaszewska

Abstract This paper aims at assessing selected approaches to measuring the effectiveness of investment in tertiary education and their applicability. It summarizes various results obtained in the research project *Methods of Measuring the Return on Investment in Higher Education*. The applied methods, include classical methods (ANOVA, Mincerian earnings function, correspondence analysis, hierarchical agglomerative clustering) as well as new ideas (application of the Wilcoxon Matched-Pairs Signed-Rank Test to determine the significance of differences in incomes before and after reaching the tertiary education). The research is based on data coming both from Polish (Social Diagnosis, Study of Human Capital) as well as German databases (Social Economic Panel, SOEP). The obtained results support the hypothesis that tertiary education influences the level of incomes. Moreover, the estimated pseudo rates of return to education

Józef Dziechciarz
Wrocław University of Economics, Poland
✉ jozef.dziechciarz@ue.wroc.pl

Marta Dziechciarz-Duda
Wrocław University of Economics, Poland
✉ marta.dziechciarz@ue.wroc.pl

Anna Król
Wrocław University of Economics, Poland
✉ anna.krol@ue.wroc.pl

Marta Targaszewska
Wrocław University of Economics, Poland
✉ marta.targaszewska@ue.wroc.pl

ARCHIVES OF DATA SCIENCE, SERIES A
KIT SCIENTIFIC PUBLISHING
Vol. 1, No. 1, S. 103–127, 2016

DOI 10.5445/KSP/1000058747/07
ISSN 2363-9881

provide the basis for the evaluation of the effectiveness of private investment in education.

1 Introduction

Since the year 1998, when the Sorbonne Declaration was signed and the European Higher Education Area was established, the development and modernization of higher education have been priorities of the policies of the European Union (EU) policies. Both, the *Bologna Declaration* and the *Lisbon Strategy*, have emphasized the following aims: improving the quality of education, building the knowledge-based society and economy, adapting the education system to the needs of the labour market, lifelong learning, and supporting the acquisition of skills to compete in a global environment. According to the strategy presented by the European Commission in 2006, the project of the Modernization Agenda for Universities (entitled *Delivering on the Modernisation Agenda for Universities: Education, Research and Innovation*) should be based on three reforms: curricula, governance and funding (cf. European Commission, 2006).

The latest EU strategy – "Europe 2020" – is yet another step of reforming the higher education system in Europe. Its main priority is to support the creation of a knowledge-based and balanced economy which favors social inclusion and cohesion. Tertiary education is one of the essential factors in achieving the main goals of this strategy. In order to define and realize all the educational aims of the strategy, the European Commission issued the Higher Education Modernisation Agenda which recommends – among other measures – increasing the number of universities' graduates, encouraging people from various social groups to undertake studies, increasing the quality of tertiary education, adjusting the curricula to labour market needs, directing higher education on financial crisis issues, as well as introducing outcome oriented funding of the universities (output-budgeting).

Moreover, the contemporary research and education market with its increasing number of students, globalization, rapid technological development, growing research costs, emergence of specialized university-independent B+R centres, and the increasing significance of commercialization and entrepreneurship poses many challenges for traditional universities and enforces their transformation (cf. Etzkowitz and Peters, 1991; Wissema, 2009; Jongbloed, 2010). In order to ensure the better quality, effectiveness and accessibility of higher education among all EU countries the shift from the traditional Humboldtian University towards the modern entrepreneurial university is essential. Conventionally functioning areas of the university – education and research – should be supplemented by other fields such as research commercialization, application for external grants, and projects as well as co-operation with industry.

A significant part of the postulated reforms in the functioning of the universities require changes in the area of funding, in particular encouraging a shift from a centralized input oriented funding mechanism towards a decentralized outcome oriented financing. All this causes the necessity of measuring the effectiveness of various aspects of higher universities' activities, including education (cf. Dziechciarz, 2011).

This paper aims at assessing selected approaches to measuring effectiveness of investment in tertiary education and their applicability, making use of the data from Polish (Social Diagnosis, Study of Human Capital), as well as German databases (SOEP). It summarizes various results obtained in the framework of the research project *Methods of Measuring the Return on Investment in Higher Education.*

2 Rate of return to education concepts

In this paper *effectiveness* refers to a relationship between higher education, resources used in education and outcomes – labour productivity and graduates' employability (cf. Aubyn et al, 2008, p. 55). One of the concepts used in measuring effectiveness in the education system is the rate of return on investment to education. The most widely and commonly used approach is the concept of private returns, measured from the point of view of individuals (students), where benefits are increased earnings and costs are foregone earnings, education fees, cost of attendance or other incidental expenses during the period of studies (cf. Psacharopoulos, 1995). The returns to education may also be measured from the social perspective. The costs are in this case the state's and the society's large spending on education and the benefits are based on productivity (cf. Psacharopoulos, 1995). Table 1 presents various types of benefits of education from both private and social perspectives.

The focus of this study is placed on measuring the effectiveness of tertiary education, defined as post-secondary education obtained at both universities and colleges.

Table 1 Classification of the benefits of education

	Private Benefits	Social Benefits
Market	employability higher earnings and savings less unemployment labor market flexibility greater mobility	higher productivity higher net tax revenue less reliance on government financial support technological development
Non-market	increased happiness better personal and family health better child cognitive development greater longevity greater satisfaction from consumption decisions	reduced crime less spread of infectious diseases lower fertility better social cohesion voter participation

See e.g. Psacharopoulos (2009), McMahon (1997).

3 Datasets

The described research is based on three datasets: German database Socio-Economic Panel Study (SOEP) (Wagner et al, 2007), and two Polish bases: Social Diagnosis (Rada Monitoringu Społecznego, 2003-2011) and the Study of Human Capital (BKL) (Bilans Kapitału Ludzkiego, 2012).

The SOEP is an annual wide-ranging representative longitudinal study on private households which started in 1984. The data provides information on households and its members and some of the many aspects include household composition, occupational biographies, employment, earnings, health and satisfaction indicators.

The Polish Social Diagnosis is a panel study investigating households and their members aged 16 and above. The project takes into account all the significant aspects of life, both the economic ones (i.e. income, material wealth, savings and financing), and the not strictly economic ones (i.e. education, medical care, problem-solving, stress, psychological well-being, lifestyle, pathologies, engagement in the arts and cultural events). The first sample was taken in the year 2000. The following study took place three years later, and since then has been repeated every two years. The database is open and may be accessed through the internet site of the panel[1].

The BKL is a labour market monitoring project carried out by the Polish Agency for Enterprise Development (Polska Agencja Rozwoju Przedsiębiorczości, PARP) in collaboration with the Jagiellonian University Krakow. In the

[1] `http://www.diagnoza.com/`

years 2010-2014 the project traced how the structure of competences changed in the labour market and sought answers to the key questions related to human capital at both the national and regional level. The project provides access to its results and gathered data without any limitations and fees. [2]

4 Various reseach approaches to measuring effectiveness of tertiary education

4.1 Application of the Mincer model in the analysis of the influence of tertiary education on the level of incomes

The first approach to measure effectiveness of tertiary education is a two step procedure. In the first step we examine the significance of the influence of education on monthly net incomes. Additionally we investigate whether factors such as sex, the class of residence, region, study major, occupation, age, tenure of employment, tenure of employment with current employer etc. significantly differentiate the income level among persons with higher education of those with lower education.

The number of observations for the analysis of the influence of education level on monthly net incomes is 9756, and 2022 for the investigation of the influence of additional factors on incomes. Only those respondents which are of working age, and who currently work (according to the variable tenure of employment with current employer), and who declared salaries at least as high as the minimum wage in the year 2009 were chosen. The research also excluded those respondents which declared extreme incomes. The list of the variables with mean standard derivation and number of observations used in the research is given in Tables 2 and 3.

In the described research one-way analysis of variance (ANOVA) is applied, except in cases of heterogeneous variances in groups of independent variables, in which the Welch test was used (cf. Proust, 2009, p. 141). ANOVA is used to examine the equality of group means for a quantitative outcome. The goal of one-way ANOVA is to verify the hypothesis that the analysed variable is influenced by independent (grouping) variables by rejecting the null hypothesis that all of the group means are equal (cf. Walesiak and Gatnar, 2012, p. 104). The application of one-way ANOVA is limited by the following assumptions:

[2] `http://en.bkl.parp.gov.pl/`

Table 2 Independent variables characteristics of the data set Social Diagnosis (2003–2011)

Variable name	Groups	Number of observations	Mean[1] [PLN]	Standard deviation
education level	Higher education	2221	2331	1013
	Post-secondary education	409	1695	548
	Secondary vocational education	2504	1761	636
	Secondary general education	789	1617	570
	Basic vocational education	3101	1637	602
	Lower secondary, primary or unfinished primary education and without education	932	1423	437
age	working mobile age (18–44 years)	1408	2286	987
	working immobile age (females 45–59, males 45–64 years)	614	2589	1031
sex	male	785	2720	1089
	female	1237	2161	891
the class of residence	big cities (100000 and more inhabitants)	924	2567	1076
	small and medium cities (less than 100000)	650	2251	2251
	villages	448	2171	2171
region [2]	central (without Warsaw)	226	2298	1003
	south (without Silesia)	150	2296	855
	east	392	2128	909
	north-west	319	2374	993
	south-west	198	2436	1075
	north	329	2476	1007
	Warsaw sub region	210	2778	1121
	Silesia	198	2386	942
tenure of employment [3]	less than 5 years	366	1964	943
	at least 5 but less than 20 years	959	2389	987
	at least 20 years	690	2579	1026
tenure of employment with current employer	5 years or less	977	2240	1021
	more than 5 years	1045	2507	982

[1] Monthly net income
[2] Warsaw sub region and Silesia were analysed separately due to higher income levels than observed in other regions in Poland (`www.wynagrodzenia.pl/dane_gus.php`, [14.11.2012])
[3] Ranges indicated by: Ustawa o promocji zatrudnienia i instytucjach rynku pracy z dnia 20 kwietnia 2004 r. [Dz. U. 2004 nr 99, poz. 1001].

the dependent variable should be normally distributed and the variance should be homogeneous in all group of independent variables (cf. Ntoumanis, 2001, pp. 73, 74).

Table 3 Independent variables characteristics of the data set Social Diagnosis (2003–2011) (cont.)

Variable name	Groups	Number of observations	Mean [1] [PLN]	Standard deviation
study major [2]	education	375	2137	826
	arts, humanities	203	2272	915
	social sciences, journalism, information sciences, economy and administration, law	717	2367	1024
	biological sciences, physics, mathematics, statistics, computer sciences	193	2565	1105
	technical sciences, production and processing, architecture and engineering	271	2631	1035
	agriculture, forestry, fishing, veterinary medicine, public health, health care, social welfare, services for population, transportation services, protection of environment and sanitary, municipal services, protection and safety	257	2434	1090
occupation [3]	parliamentarians, high officials and managers	223	2984	1078
	specialists	1080	2399	947
	technicians and other mid-level staff	336	2308	1028
	office workers	156	1958	8778
	personal services staff and salesmen, farmers, gardeners, foresters, fishermen, industry workers, craftsmen, operators and mechanics for machines, simple work staff, armed forces	166	2054	986

[1] Monthly net income
[2] Classification indicated by: Rozporządzenie Rady Ministrów w sprawie Polskiej Klasyfikacji Edukacji z dnia 6 maja 2003 r. [Dz. U. 2003 nr 98, poz. 895].
[3] Classification indicated by: Rozporządzenie Ministra Pracy i Polityki Społecznej w sprawie klasyfikacji zawodów i specjalności na potrzeby rynku pracy oraz zakresu jej stosowania z dnia 27 kwietnia 2010 r. [Dz. U. 2010 nr 82, poz. 537].

The research was conducted for the data from the Social Diagnosis data set in the year 2009, and all hypotheses (influence of an independent variable on income) were verified at the 95% confidence level. The test statistics and significance levels for one-way ANOVA or Welch test are presented in Table 4.

The results of the analysis show that the level of education significantly influences the monthly net income. The highest incomes were characteristic for persons with tertiary education degree, and lowest for persons with at most a lower-secondary education level. Moreover, all independent variables

Table 4 Test of influence of independent variable on income for 2009 of the data set Social Diagnosis

Variable name	Statistic	Significance – ANOVA/ strong tests for means equality
education level	245.500	0.000
age	39.193	0.000
sex	145.514	0.000
the class of residence	31.126	0.000
region	8.774	0.000
tenure of employment	46.358	0.000
tenure of employment with current employer	35.695	0.000
major	11.027	0.000
occupation	31.317	0.000

significantly (at the level 0.05 for the post-hoc Games-Howell test (cf. Morgan et al (2004, p. 152), Field (2005, p. 341)) differentiate personal monthly net income of persons with higher education degree, where the average monthly income is 2331.16 PLN. The results of research also shows that females achieve lower (on average of 550 PLN) wages than males. Lower incomes are specific for persons of working mobile age. This could be explained by the lower experience of those persons.

There is also a significant difference between the wages of persons living in big cities (population $\geq$ 100000) and the persons living in small and medium cities (population $<$ 100000) or villages, where the wages of persons from the first two categories are the highest. In case of the variable "region" high income is characteristic for the respondents from the Warsaw sub-region, and the lowest income is observed for the Eastern provinces (Lublin, Podkarpackie, Podlaskie, Świętokrzyskie).

Both, "tenure of employment" and "tenure of employment with current employer", significantly influence the achieved income - persons with higher experience (more than 5 years) earn more money. The last two variables – "study major" and "occupation" - are closely related. Our research shows that the graduates of educational studies, humanities or art studies obtain lower incomes than graduates of technical and theoretical science studies. Moreover, the average incomes in the group of social sciences, journalism and information sciences, economy, administration and law studies are similar to the group which contains studies such as agriculture, forestry, fishing, veterinary medicine, public health, health care, social welfare, services for population, transportation services, protection of environment, sanitary municipal services, protection, and safety. The level of monthly net income is also influenced by occupation, where the highest incomes (800 PLN higher than for other professions) are specific to parliamentarians, high officials and managers. Similar earnings are characteristic for technicians, mid-level staff and specialists. For more details on the result of this analysis see Targaszewska (2013).

These results were used to support the process of cmodel selection in the second step of the research - the estimation of the private pseudo rate of return to education with Mincer's earning function. Additionally, the analysis of the dynamics of the influence of the education level on wages was performed. The empirical research for Germany was based on the SOEP database (years 1995, 2000, 2005, 2010) and for Poland on the Social Diagnosis data set (years 2003, 2005, 2007, 2009, 2011). The description of variables for the SOEP data set and its summary statistics are presented in Table 5, whereas the presentation of the variables from the Social Diagnosis data set is given in Table 6. For the sake of conciseness, the summary statistics for the SOEP data set are presented only for the year 2010 (the most recent of the analyzed years) and for the Social Diagnosis data set only for the year 2009 (as in Table 4). Similarly, the various specifications of the estimated models are presented in detail in Tables 7 and 8 only for the selected years (the SOEP data set in year 2010 and the Social Diagnosis data set in year 2009 accordingly), and Tables 9 and 10 present only the most important results for each year of the study.

To analyze the dynamics of influence of education level on the wages the commonly applied Mincerian earnings function (cf. Mincer, 1958, 1974) was used in each year separately:

$$lnEAR_i = X_i^T \beta + \varepsilon_i, \tag{1}$$

where EAR - earnings, X - vector of variables influencing wages, β - vector of unknown parameters, ε - error term. The elements of X describe education (represented by number of years of education or dummies for level of education, the latter providing the estimation of the pseudo rate of return to education) and professional experience, as well as auxiliary characteristics such as gender, region, place of work, position etc. As the dependent variable real hourly gross earnings for SOEP data set and real monthly net earnings for Social Diagnosis data set were used. Adjustment for inflation led to creation of new variables: *RHGEAR* and *RMNEAR* accordingly. Afterwards the inflation-adjusted earnings were transformed to natural logarithms. The log-linear functional form proved to be correct in many previous studies (cf. Heckman et al, 2003). In addition, a Box-Cox transformation which allows to choose between linear ($\alpha = 1$) and log-linear ($\alpha = 0$) specification was used (cf. Box and Cox, 1964):

Table 5 Description of variables for the SOEP data set and summary statistics in year 2010

Variable name	Description	Mean	Standard deviation	Distribution of categories (%)
HGEAR	Hourly gross earnings in [EUR]	16.80	15.01	
YOET	Years of education and training	12.82	2.75	
AGE	Age	43.70	12.52	
PWE	Potential work experience (*PWE*=*AGE*-*YOET*-6)	25.60	12.18	
HEDU	Higher education			
	1 if obtained higher education diploma			26.60
	0 otherwise			73.40
MEDU	Secondary education			
	1 if obtained secondary education diploma			98.08
	0 otherwise			1.92
SEN	Tenure of employment with current employer	11.53	10.57	
FEM	Gender			
	1 if female			48.90
	0 otherwise			51.10
TYPE	Work position type			
	APP (trainee)			12.74
	SPEC (specialist)			12.17
	PROF (freelancer/professional)			44.69
	MAN (manager)			6.74
	OTHER (other)			23.66
SIZE	Size of current employer			
	SMALL (less than 20 employees)			31.66
	MEDIUM (20 - 2000 employees)			47.29
	LARGE (more than 2000 employees)			21.05

$$B(EAR_i,\alpha)=\begin{cases}\frac{EAR_i^{\alpha}-1}{\alpha} & \text{for } \alpha\neq 0\\ \ln EAR_i & \text{for } \alpha=0\end{cases}. \tag{2}$$

For all tested specifications the parameter α was close to 0 indicating the better fit of the models resulting from the log-linear transformation. Fig. 1 presents the results of searching the parameter α which maximizes the logarithm of the likelihood function for the model specification SOEP4 in Table 7. The horizontal line in Fig. 1 marks the 95% confidence interval for parameter α. Its lower boundary (0.1051), center (0.1213) and upper boundary (0.1353) are shown by vertical lines.

Table 6 Description of variables for the Social Diagnosis data set and summary statistics in year 2009

Variable name	Description	Mean	Standard deviation	Distribution of categories (%)
MNEAR	Monthly net earnings in [PLN]	1350.00	707.72	
HE	Higher education			
	1 for tertiary education			10.83
	0 otherwise			89.17
ME	Secondary education			
	1 for secondary education			36.63
	0 otherwise			63.37
YOE	Years of education	11.54	3.31	
AGE	Age	48.77	18.01	
EXP	Professional experience years	22.36	13.91	
WSEC	Work sector			
	public (*PUB*)			14.21
	private (*PRIV*)			24.08
	own business (*ENT*)			3.20
	other (*OTH*)			58.51
CTYPE	The type of residence			
	BCITY (more than 100 thousand occupants)			23.92
	MCITY (less than 100 thousand occupants)			32.90
	VIL (villages)			43.18
F	Gender			
	1 if female			54.79
	0 otherwise			45.21
EAST	Geographical localization[1]			
	1 if eastern Poland			26.52
	0 otherwise			73.48

[1]Eastern Poland includes the Lubelskie, Podkarpackie, Podlaskie, Świętokrzyskie and Warmińsko-Mazurskie regions in accordance with the division incorporated in the European Operational Programme Development of Eastern Poland. These provinces are considered to have lower living standards, a lower dynamic of economic development, poorly developed and inadequate transport infrastructure and insufficient growth factors, which might be reflected in the earnings of their residents.

Since heteroscedasticity of the error term was detected in the large majority of cases (using White's test (cf. White, 1980)), for model estimation the weighted least squares method was applied. For a more detailed analysis and additional information see Król (2014).

Table 7 presents estimation results of five different specifications of Mincer's model in the year 2010 based on the SOEP data set (dependent variable

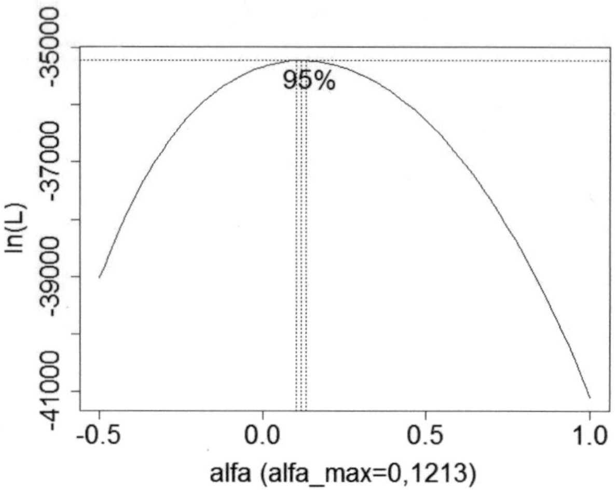

Fig. 1 Values of the logarithm of likelihood function for various values of α parameter for specification (SOEP4).

ln(RHGEAR)), starting from the simplest classic Mincer model SOEP1 describing earnings by the number of years of education and experience to the most complex model SOEP5. Model SOEP5 was chosen for further interpretation, since it has the highest goodness-of-fit measure and the lowest AIC information criterion, moreover all its variables are statistically significant. The interpretation of results of estimation of the model SOEP5 shows that in the year 2010 the Germans with higher education could earn about 25% more in comparison to similar (in terms of gender, work experience, work type, size of the company etc.) persons. Women in Germany earned in 2010 on average about 15% less than men on similar work positions and with similar professional experience. The professionals and freelancers in the year 2010 could earn about 31% more, the managers about 33% more and the trainees about 31% less than regular employees (specialists and other employees). This result shows that labour market requirements extend beyond simple higher education diploma, and that additional qualifications and skills are also important. The differences in earnings in big and small firms may also be observed: employees of small companies in the year 2010 earned about 24% less and in medium companies about 10% less than employees in big corporations, *ceteris paribus*.

Table 8 presents estimation results of five specifications of Mincer's model in the year 2009 based on the model Social Diagnosis data set (dependent

Table 7 Estimation results of five different specifications of Mincer's model (year 2010) based on the SOEP data set (dependent variable *ln(RHGEAR)*)

	SOEP1	SOEP2	SOEP3	SOEP4	SOEP5
constant	0.8636***	1.839***	1.518***	2.207***	2.090***
	(0.03461)	(0.02436)	(0.07071)	(0.02290)	(0.06951)
YOET	0.08217***				
	(0.001909)				
PWE	0.05246***	0.05220***	0.05225***	0.02572***	0.02577***
	(0.001943)	(0.002006)	(0.002001)	(0.001813)	(0.001803)
PWE^2	−0.0008456***	−0.0008823***	−0.0008815***	−0.0004860***	−0.0004867***
	(3.867e-05)	(3.916e-05)	(3.912e-05)	(3.522e-05)	(3.508e-05)
HEDU		0.4192***	0.4148***	0.2231***	0.2229***
		(0.01225)	(0.01226)	(0.01046)	(0.01041)
MEDU			0.3236***		0.1181*
			(0.06703)		(0.06579)
SEN				0.02148***	0.02151***
				(0.001403)	(0.001396)
SEN^2				−0.0003143***	−0.0003161***
				(3.659e-05)	(3.644e-05)
FEM				−0.1679***	−0.1684***
				(0.008819)	(0.008774)
APP				−0.3813***	−0.3749***
				(0.02136)	(0.02131)
PROF				0.2694***	0.2682***
				(0.01036)	(0.01031)
MAN				0.2893***	0.2881***
				(0.01771)	(0.01765)
SMALL				−0.2836***	−0.2847***
				(0.01369)	(0.01361)
MEDIUM				−0.1084***	−0.1081***
				(0.01028)	(0.01021)
n	9534	9534	9534	8787	8787
$\bar{R}^2$	0.2219	0.1711	0.1731	0.4153	0.4172
AIC	41081.68	40710.35	40745.02	38393.00	38289.02

Significance levels ('***' = 0.01; '**' = 0.05; '*' = 0.1), n = number of observations , $\bar{R}^2$ adjusted R^2 and AIC = Akaike Information Criterion.

variable *ln(RMNEAR)*). Again the most complex specification (SD5) is taken for interpretation and further research. The premium for higher education in Poland in the year 2009 was about 29%. Note that in the Social Diagnosis data set the values of variables *HE* and *ME* for the persons with higher education are 1, whereas in the SOEP data set for the persons with higher education *HEDU*=1 and *MEDU*=0. The difference in earnings between males and females in Poland is bigger than in Germany. In the analysed period women earned on average about 19% less than men doing similar work in much the same work

Table 8 Estimation results of five different specifications of Mincer's model (year 2009) based on the Social Diagnosis data set (dependent variable *ln(RMNEAR)*)

	(SD1)	(SD2)	(SD3)	(SD4)	(SD5)
constant	6.133***	6.873***	6.738***	6.916***	6.795***
	(0.01718)	(0.01060)	(0.01213)	(0.01385)	(0.01437)
YOE	0.06484***				
	(0.001002)				
EXP	0.01195***	0.01404***	0.01354***	0.01105***	0.01288***
	(0.0009639)	(0.0009927)	(0.0009839)	(0.0008900)	(0.0008705)
EXP^2	−0.0001213***	−0.0002492***	−0.0002102***	−8.475e-05***	−0.0001101***
	(1.964e-05)	(2.104e-05)	(2.004e-05)	(1.867e-05)	(1.793e-05)
HE		0.4807***	0.6024***	0.3470***	0.4699***
		(0.009995)	(0.01068)	(0.009400)	(0.01008)
ME			0.2887***		0.2116***
			(0.008038)		(0.007129)
MCITY				−0.08492***	−0.06582***
				(0.008162)	(0.008016)
VIL				−0.2194***	−0.1637***
				(0.008138)	(0.008188)
PUB				0.4295***	0.3852***
				(0.009388)	(0.009046)
PRIV				0.3909***	0.3662***
				(0.008321)	(0.008243)
ENT				0.5190***	0.4745***
				(0.01904)	(0.01881)
EAST				−0.08472***	−0.08729***
				(0.007100)	(0.006818)
F				−0.1933***	−0.2138***
				(0.006624)	(0.006485)
n	18417	18426	18426	18370	18370
$\bar{R}^2$	0.1904	0.1182	0.1642	0.3330	0.3660
AIC	77441.29	76479.20	75850.83	77973.57	77847.24

Significance levels ('***' = 0.01; '**' = 0.05; '*' = 0.1), n = number of observations , $\bar{R}^2$ adjusted R^2 and AIC = Akaike Information Criterion.

place. The influence of residence on the level of wages was significant as well. Eastern Polish regions, which are considered to have lower living standards, lower dynamic of economic development and insufficient growth factors, show significantly lower earnings. In comparison to Central and West Poland the people from eastern provinces earned about 8% less, *ceteris paribus*. Moreover, the residents of small and medium cities earn about 6% and residents of villages about 15% less than the inhabitants of big cities.

Tables 9 and 10 present the final estimated models for the years 1995–2010 for the SOEP data set and for the years 2003 – 2011 for the Social Diagnosis data

Table 9 Estimation results of Mincer's model in years 1995, 2000, 2005, 2010 based on the SOEP data set (dependent variable *ln(RHGEAR)*)

	(SOEP1995)	(SOEP2000)	(SOEP2005)	(SOEP2010)
constant	2.421**	2.502**	2.243**	2.090**
PWE	0.02106**	0.001609**	0.03010**	0.02577**
PWE^2	−0.0004066**	−7.616e-07**	−0.0005706**	−0.0004867**
HEDU	0.09978**	0.1385**	0.1826**	0.2229**
MEDU	−0.05301	0.04387	0.02129	0.1181*
SEN	0.01831**	0.02322**	0.02000**	0.02151**
SEN^2	−0.0004006**	−0.0004539**	−0.0003263**	−0.0003161**
FEM	−0.1897**	−0.1899**	−0.1789**	−0.1684**
APP	−0.3114**	−0.4339**	−0.4124**	−0.3749**
PROF	0.2190**	0.2301**	0.2636**	0.2682**
MAN	0.1461**	0.1861**	0.2544**	0.2881**
SMALL	−0.2908**	−0.2744**	−0.2894**	−0.2847**
MEDIUM	−0.1080**	−0.1101**	−0.1202**	−0.1081**
n	7018	12423	9697	8787
$\bar{R}^2$	0.3365	0.3474	0.4088	0.4172
AIC	29487.92	53348.74	42004.25	38289.02

Significance levels ('***' = 0.01; '**' = 0.05; '*' = 0.1), n = number of observations , $\bar{R}^2$ adjusted R^2 and AIC = Akaike Information Criterion.

Table 10 Estimation results of Mincer's model in years 2003, 2005, 2007, 2009, 2011 based on the Social Diagnosis data set (dependent variable *ln(RMNEAR)*)

	(SD2003)	(SD2005)	(SD2007)	(SD2009)	(SD2011)
constant	6.010***	6.067***	6.131***	6.795***	6.677***
HE	0.5349***	0.4349***	0.4805***	0.4699***	0.4879***
ME	0.2171***	0.2226***	0.2286***	0.2116***	0.2353***
MCITY	−0.09271***	−0.07576***	−0.07770***	−0.06582***	−0.04613***
VIL	−0.2105***	−0.1382***	−0.1452***	−0.1637***	−0.1384***
PUB	0.4884***	0.4087***	0.4276***	0.3852***	0.4390***
PRIV	0.4486***	0.3173***	0.3862***	0.3662***	0.3959***
ENT	0.6216***	0.4372***	0.5128***	0.4745***	0.4716***
EAST	−0.05617***	−0.05287***	−0.03427***	−0.08729***	−0.07752***
F	−0.2324***	−0.1882***	−0.2030***	−0.2138***	−0.2219***
n	6707	5802	9205	18370	18661
$\bar{R}^2$	0.3497	0.3028	0.3081	0.3660	0.3545
AIC	29570.07	24745.31	39952.75	77973.57	80272.48

Significance levels ('***' = 0.01; '**' = 0.05; '*' = 0.1), n = number of observations , $\bar{R}^2$ adjusted R^2 and AIC = Akaike Information Criterion.

set. The obtained results allow for the evaluation of the dynamics of influence of higher education and other factors on the earnings in Germany and Poland accordingly.

In Germany in the last 15 years we observe a quite stable increase in the value of the premium for higher education (from 11% to 25%). The auxiliary factors whose influence changed the most are the ones connected with the type of work. For example, the premium for managers increased from about 15% in the year 1995 to about 33% in the year 2010. Another interesting trend observed in the analyzed period is the slight decrease of gender-related work discrimination (from about 17% to about 15%).

The analysis of Polish data shows the stabilization of the influence of tertiary education on the level of earnings. In the years 2005 – 2011 the premium for higher education oscillates around the level of 30%. Similarly, the earnings of women in the analyzed period remain lower than those of men of about 20%. There is a slight improvement in the reduction of regional differences. The difference in earnings of the residents of villages in comparison to the inhabitants of big cities decreased from about 19% in 2005 to about 15% in 2011.

4.2 Determination of the significance of differences in incomes before and after reaching the higher education

Another part of our research was to check if there is a significant difference in incomes before and after reaching a higher education degree and, in addition, to measure the rate of return to education in both groups (cf. Targaszewska, 2014). To achieve those goals the Wilcoxon matched-pairs signed-rank test and the classical Mincerian function were applied. The Wilcoxon matched-pairs signed-rank test is non-parametric test used to compare two paired (dependent) samples – each observation of the first sample has a unique connection with an observation in the second sample. The null hypothesis states the equality of median difference in paired observations. This means: samples have identical distributions (cf. Jackson, 2011, pp. 266, 267). The research was based on the group of respondents which fulfilled the following conditions:

- They participated in the SD-project in the years since 2003,
- they declared a lower than tertiary level of education,

- and they declared in 2011 to have a higher level of education than in the previous studies.

For both groups the rate of return to education is measured and compared (Targaszewska, 2014). Because of the nature of the data some assumptions were made. Firstly, in a situation where over the years, some respondents changed their level of education more than once, the research included only the most recent change. Secondly, the variable denoting years of experience in 2011 was estimated. Experience in 2011 is equal to experience in 2009 plus two years. Lastly, incomes were corrected by the inflation indicator (with the base year 2003). Moreover, the cases with incomes under the minimum wages in each year were removed from the research. Finally 152 cases were taken into account. The p-value for the executed test was almost 0.0. This allows to reject the null hypothesis (of equality of median difference in paired observations) which means that there is a significant difference in incomes between groups. Wages of persons with a higher education degree are on average higher by 773 PLN in comparison to persons without this kind of education. 2453 [PLN] and 3226 [PLN] are the means "before" and "after" reaching the level of higher education, respectively.

Subsequently, the rate of return to education and rate of return to experience for both groups "before" and "after" were estimated by the classical Mincer model (cf. Mincer, 1974):

$$\log(Y) = \alpha + \rho s + \beta_0 x + \beta_1 x^2 + \xi, \tag{3}$$

where Y is earnings, s is schooling level or years of study, x is work experience. The parameter ρ can be interpreted as the average private rate of return to schooling, β is related to the financial return to experience, and α is related to initial earnings capacity (cf. Polachek, 2008). The estimated models are presented in Table 11.

In the first model for the group "before" the parameter p is not significant at the 5% significance level. It seems that for persons without higher education degree the most important variable is experience. After rejecting the variable "years of study" the model was estimated once again. From the new model for group "before" one can conclude that the rate of return to experience, after 10 years of working is nearly 1.4%. For the group "after" each of the parameters is significant. The rate of return to education is about 6.6% and rate of return to experience after 10 years of working is about 1.1%.

Table 11 Parameter estimates of Mincer's model based on the data set Social Diagnosis (2003–2011)

	Group					
	before		before without years of study		after	
Variable	Coefficient	p-value	Coefficient	p-value	Coefficient	p-value
Constant	7.474	0.000	7.458	0.000	6.584	0.000
Years of study (s)	-0.001	0.943	–	–	0.066	0.002
Experience (x)	0.034	0.000	0.034	0.000	0.031	0.001
Quadratic experience (x^2)	-0.001	0.006	-0.001	0.006	-0.001	0.017
n	152		152		152	
$\bar{R}^2$	0.1465		0.1522		0.1792	
AIC	158.39		156.36		143.84	

n = number of observations , $\bar{R}^2$ adjusted R^2 and AIC = Akaike Information Criterion.

4.3 Examination of non-monetary benefits of tertiary education

Our next research step for measuring the effectiveness of education was to capture the intangible benefits of higher education, particularly non-monetary private and social rates of return on investment in education. Empirical studies were carried out on data from the Social Diagnosis 2011 data set. As shown in Table 1, non-monetary returns are an important part of the benefits of education. It is commonly believed that better educated people have a better life. This general opinion can be empirically confirmed in two ways. Firstly, by people's personal experience and, secondly by the statements of the respondents concerning their life quality perception and expectations along with their level of education. Fig. 2 visualizes the output of the correspondence analysis performed on this data. A comprehensive description of the algorithm of correspondence analysis, computational details, and its applications can be found in the classic text by Greenacre (1984). Fig. 2 shows the coincidence of the respondents' education level with a subjective evaluation of the happiness with his/her life in the last years.

The position of higher education in Fig. 2 is close to the most positive assessment of one's life in the last year. The percent of total inertia described in the two first dimensions is almost 100%. In Figs. 2 and 3, the first dimension describes about 98% of inertia. In Fig. 2, 'very happy' is next to 'high education', and further to the right the 'education level' is lower, and the 'assessment of life in the last year' is also getting worse. Basic education is near to negative assessment. The conclusion is that better education is associated with a more

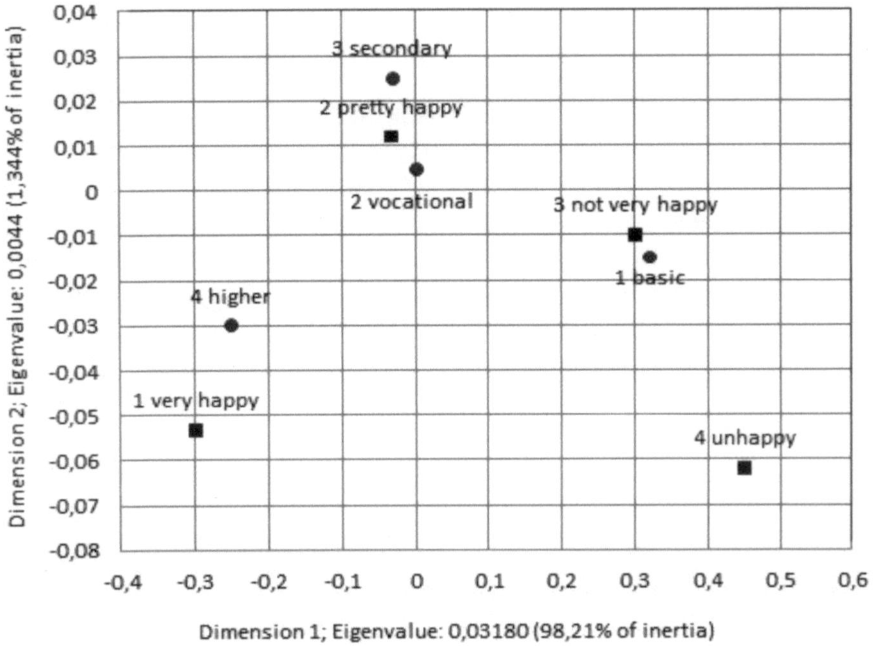

Fig. 2 Correspondence analysis of education variable and assessment of life in the last year, the Social Diagnosis 2011, sample size: 26332. Education Levels: 1 – basic education, 2 – vocational, 3 – secondary and 4 – higher. Frequency of health problems: 1 – often, 2 – sometimes and 3 – never.

positive perception of the past. But when respondents were asked to name the three most important conditions for a successful and wonderful life, the first five positions were health, children, happy marriage, work and money. The education level was mentioned somewhere between the 13th and 10th place out of 14 possible places (higher position for better educated respondents). This surprising phenomenon can be explained by the association of education with higher earnings and better work.

Private non-monetary returns of tertiary education include the impact of education on personal health, the ability to enjoy leisure and the capacity to make personal choices. Obviously, education tends to improve income which affects health positively. People with a higher education level are more aware of healthy behaviour and demonstrate more tendencies to seek treatment when needed. More results of the analysis of non-monetary benefits of tertiary education can be found in Dziechciarz-Duda and Król (2013).

According to the WHO Regional Office for Europe (2012), the male population (age of 30) with higher education will live on average another 48.5 years. While the male population (age of 30) with primary education will live on average another 36.5 years and for secondary education another 43 years. For women, life expectancy is in general higher: for better educated women on average 83.2 years and for the least educated women 5 years shorter. Moreover, differences in the risk of death related to the educational level are greater in the case of men than women for all causes of death (except cardiovascular diseases). Death rates from all main causes tend to be lower among people with higher education levels. All diseases contribute to shortening the lives of less-educated people more than the lives of better educated individuals. The cause that is most responsible for shortening the lives of less-educated people when compared with better educated individuals are cardiovascular diseases, external causes and cancer. Numerous research results confirm that higher education contributes to increased longevity and better health in terms of severe and fatal diseases, partly through the increased earnings that enable the purchase of better health care and a better diet (cf. WHO Regional Office for Europe, 2012).

Fig. 3 visualizes the output of the correspondence analysis for the level of education of Polish respondents in 2011 with the subjective evaluation of self-well-being expressed as the frequency of health problems that hinder a positive perception of the quality of life. The results support the hypothesis of the positive impact of education on personal health. In Fig. 3, "often" is next to "basic education" and further, to the right the level of education is growing and the assessment of health is better.

4.4 Analysis of employment status and professional profiles of universities graduates

The goal of our last research step was to analyse the professional situation of young people with tertiary education. For this purpose a hierarchical classification method (Ward) was applied to the data from the Study of Human Capital in Poland 2012, and 8 homogenous classes of university graduates were distinguished based on the dendrogram. The analysis of the characteristics of each class is a valuable source of information about the factors that have an impact on the level of unemployment in this group. The following variables describing the situation of graduates on the labour market were used: profes-

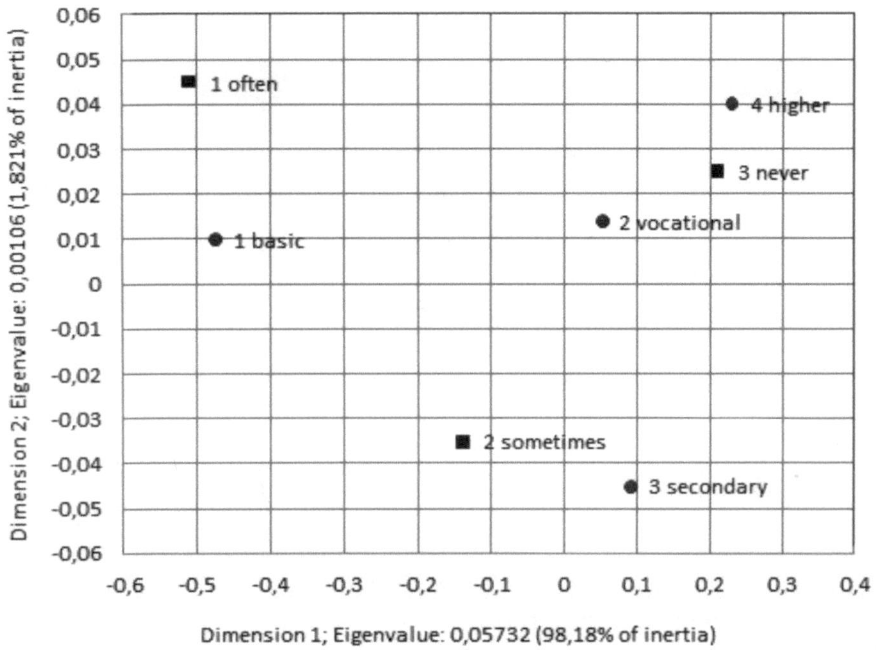

Fig. 3 Correspondence analysis of education and the frequency of health problems, the Social Diagnosis 2011, sample size: 26332. Education Levels: 1 – basic education, 2 – vocational, 3 – secondary and 4 – higher. Frequency of health problems: 1 – often, 2 – sometimes and 3 – never.

sional status (full-time job, part-time job, unemployed, housework, etc.), the type of university (private, public), the type (full-time studies, evening studies, extramural studies), the level of studies (bachelor, engineer, master, etc.) as well as the average level of net income.

The analysis of the professional situation and characteristics of the graduates in separate classes allowed for the assessment of how well the representatives of each group cope with the labour market challenges (see Fig. 4). The worst groups, in terms of the percentage of employed and earnings level, were the young, out-of-work people with a bachelor's degree, graduated from private universities (class 5), young people without work and experience (class 8), as well as young people from small towns and villages, graduated from agriculture and service studies (class 2). The level of employment and earnings in class 3 (teachers and humanistic studies graduates) is similar to the average in the whole population. Whereas the situation of the classes 1 (well-paid engineers),

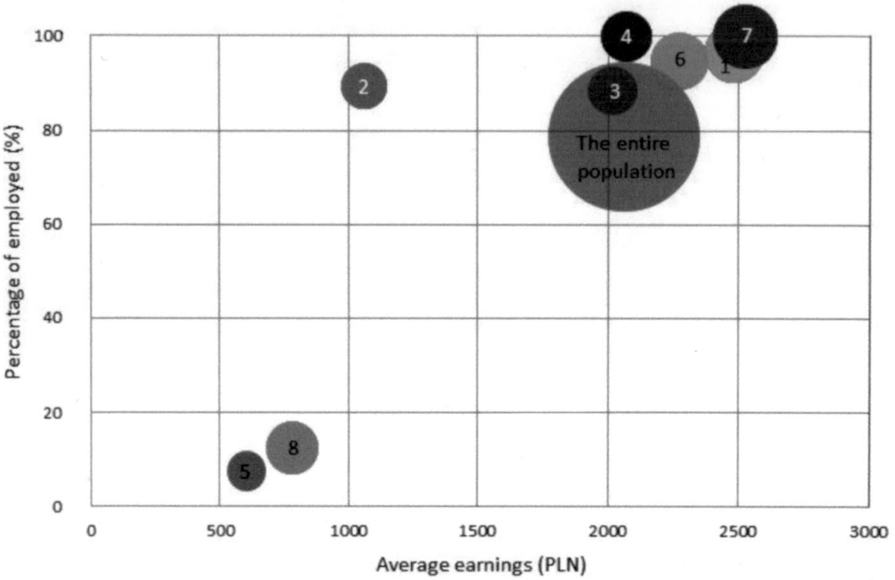

Fig. 4 Assessment of groups according to the percentage of employed and earnings, Study of Human Capital in Poland 2012.

4 (employed economists with significant experience), 6 (working with a bachelor's degree, graduated from private universities) and 7 (entrepreneurial with a master's degree) is significantly better. Comprehensive results of the research can be found in Dziechciarz-Duda and Przybysz (2014).

These results show that graduates with a bachelor degree are in less favorable condition compared to graduates with a master degree. The percentage of employment in the group of bachelor degree holders is only 63,8%, whereas for graduates with master degrees it increases to 80,7%. A similar situation may be observed for graduates of technical universities – the employment rate of undergraduate engineers is 76,5%, while in the group of engineers with master degrees it is 85,3%. The graduates of engineering and technical studies, as well as mathematics, statistics, physics and medicine grads occupy the strongest position on the labour market. The students of the most popular majors (economics, pedagogics and social studies) face an employment rate of about 80% and an unemployment rate of almost 15%.

5 Final remarks

This paper summarizes various results obtained in the framework of the research project *Methods of Measuring the Return on Investment in Higher Education*. The goal of the project was to analyse the problem of measuring the effectiveness of investment into higher education in its various forms. The research approaches, included classical methods (ANOVA, Mincerian earnings function, correspondence analysis, hierarchical agglomerative clustering), as well as new ideas (application of Wilcoxon Matched-Pairs Signed-Rank Test to determine the significance of differences in incomes before and after reaching higher education). All obtained results support the hypothesis that higher education influences the level of income. Moreover, the estimated pseudo rates of return to education provide the basis for the evaluation of the effectiveness of private investment in education.

Acknowledgements The study was conducted in the framework of the research project entitled *Rate of Return Measurement Methods in Higher Education* (*Metody pomiaru stopy zwrotu z inwestycji na edukację w szkołach wyższych*). The project has been financed by the National Science Centre on the basis of decision no. DEC-2011/01/B/HS4/02328.

References

Aubyn M, Pina A, Garcia F, Pai J (2008) Study on the efficiency and effectiveness of public spending on tertiary education. http://ec.europa.eu/economy_finance/publications/publication16267_en.pdf, accessed: 2014-12-30

Bilans Kapitału Ludzkiego (2012) http://bkl.parp.gov.pl, accessed: 2013-06-03

Box GEP, Cox DR (1964) An analysis of transformations. Journal of the Royal Statistical Society Series B (Methodological) 26(2):211–252

Dziechciarz J (2011) On rate of return measurement in education. Econometrics 31:49–66

Dziechciarz-Duda M, Król A (2013) On non-monetary benefits of tertiary education. Econometrics 41(3):78–94

Dziechciarz-Duda M, Przybysz K (2014) Wykształcenie a potrzeby rynku pracy. Klasyfikacja absolwentów wyższych uczelni. Taksonomia Klasyfikacja i analiza danych – teoria i zastosowania 22(327):303–312

Etzkowitz H, Peters LS (1991) Profiting from knowledge: Organisational innovations and the evolution of academic norms. Minerva 29(2):133–166, DOI 10.1007/BF01096406

European Commission (2006) Communication from the Commission to the Council and the European Parliament. `http://eur-lex.europa.eu/LexUriServ/LexUriServ.do?uri=COM:2006:0208:FIN:en:PDF`, accessed: 2013-12-03

Field A (2005) Discovering Statistics Using SPSS. SAGE, London

Greenacre M (1984) Theory and applications of correspondence analysis. Academic Press, New York

Heckman J, Lochner L, Todd P (2003) Fifty Years of Mincer Earnings Regressions. NBER Working Papers (9732), National Bureau of Economic Research

Jackson S (2011) Research Methods and Statistics: A Critical Thinking Approach. Wadsworth Publishing Company, Belmont

Jongbloed B (2010) Funding Higher Education: A View across Europe. University of Twente, Enschede

Król A (2014) An analysis of the dynamics of higher education's influence on the level of wages. Econometrics 43(1):60–73

McMahon WW (1997) Recent advances in measuring the social and individual benefits of education. International Journal of Education Research 27(6):447–531, DOI 10.1016/S0883-0355(97)00047-5

Mincer J (1958) Investment in human capital and personal income distribution. Journal of Political Economy 66(4):281–302

Mincer J (1974) Schooling, Experience and Earnings. Columbia University Press, New York

Morgan GA, Leech NL, Gloeckner GW, Barrett KC (2004) SPSS for Introductory Statistics: Use and Interpretation. Lawrence Erlbaum Associates, Mahwah

Ntoumanis N (2001) A Step-by-Step Guide to SPSS for Sport and Exercise Studies. Routledge, London

Polachek SW (2008) Earnings over the life cycle: The mincer earnings function and its applications. Foundations and Trends in Microeconomics 4(3):165–272, DOI 10.1561/0700000018

Proust M (2009) Statistics and Graphics Guide. SAS, Cary

Psacharopoulos G (1995) The Profitability of Investment in Education: Concepts and Methods. HCO Working Papers 63

Psacharopoulos G (2009) Returns to Investment in Higher Education. A European Survey. Center for Higher Education Policy Studies, Enschede

Rada Monitoringu Społecznego (2003-2011) Diagnoza Społeczna: zintegrowana baza danych, 2003-2011. www.diagnoza.com, accessed: 2012-08-04

Targaszewska M (2013) An attempt at identification sources of variation in monthly net incomes among persons with tertiary education. Econometrics 39(1):210–220

Targaszewska M (2014) An attempt at measuring the effectiveness of higher education in Poland. Econometrics 43(1):50–59

Wagner GG, Frick JR, Schupp J (2007) The German Socio-Economic Panel Study (SOEP) - scope, evolution and enhancements. Schmollers Jahrbuch 127(1):139–169

Walesiak M, Gatnar E (2012) Statystyczna analiza danych z wykorzystaniem programu R. PWN, Warszawa

White H (1980) A heteroskedasticity-consistent covariance matrix estimator and a direct test for heteroskedasticity. Econometrica 48(4):817–838

WHO Regional Office for Europe (2012) Social inequalities in health in Poland. http://www.euro.who.int/__data/assets/pdf_file/0008/177875/E96720.pdf, accessed: 2013-09-17

Wissema JG (2009) Towards the Third Generation University. Managing the University in Transition. EE Publishing, Cheltenham

Statistical Simulation of a Multi-Phase Tool Machining a Multi-Phase Workpiece

Swetlana Herbrandt, Uwe Ligges, Manuel Ferreira, Michael Kansteiner, and Claus Weihs

Abstract The continuing development of the multi-phase material concrete leads to an increased demand for the optimization of diamond impregnated tools. Because of high initial investment costs for diamond tools, not only the reduction of processing time, but also the reduction of tool wear is in the focus of interest. While some parameters like cutting speed can be controlled, other important parameters like the number of cutting diamonds are beyond our influence. To manage this randomness, simulation models for diamond and segment grinding are developed. In this work we will present two models for a segment grinding simulation. The first model is an extension of the simulation model proposed by Raabe et al. (2011) for single diamond scratching on basalt. Beside the goodness-of-fit, the simulation time is an essential factor in the development and choice of simulation models. The difficulties encountered while extending this model are discussed and we provide a solution to accelerate the workpiece simulation. In order to achieve a further reduction of simulation time, a second model is introduced under the assumption of pyramidal shaped diamonds. The

Swetlana Herbrandt · Uwe Ligges · Claus Weihs
TU Dortmund, Department of Statistics,
✉ [herbrandt, ligges, weihs]@statistik.tu-dortmund.de
Manuel Ferreira
TU Dortmund, Institute of Materials Engineering,
✉ manuel.ferreira@tu-dortmund.de
Michael Kansteiner
TU Dortmund, Institute of Machining Technology,
✉ michael.kansteiner@tu-dortmund.de

ARCHIVES OF DATA SCIENCE, SERIES A
KIT SCIENTIFIC PUBLISHING
Vol. 1, No. 1, S. 129–155, 2016

DOI 10.5445/KSP/1000058747/08
ISSN 2363-9881

simulation results are compared with single diamond experimental data and a feasibility study is performed for the segment setup.

1 Introduction

The machining of mineral subsoil is daily routine at building sites. In many cases the machined material is concrete and the preferred tool for trepanning is a diamond impregnated drill because of the diamond‘s cutting properties. Since these tools are in general not adapted to particular situations, under certain circumstances the tool wear can be much higher than expected and would therefore lead to an earlier need for replacement. Hence our main target is the understanding and optimization of the machining process with simultaneous reduction of tool wear. The difficulty of this task is in the complexity of the process in conjunction with the problem that many variables affecting the machining process can not be directly influenced and are even difficult to observe. The simulation should enable the control of these parameters and offer the possibility to conduct as many simulated experiments as necessary to find optimal settings for tool production and the machining process.

In the last twenty years many different models for the simulation of forces, material removal and temperature in grinding processes were presented (Brinksmeier et al, 2006). The model categories range from heuristic and empirical to physical models, while high performance computers allow for the computation of models with resolution degrees from macroscopic to microscopic. The considered areas of application are as various as the models due to the versatile usability of diamond impregnated tools, the diverse characteristics of machined materials, and the multiple kinds of machining processes. The state of the art method for simulations of engineering applications like grinding or sawing is the finite elements approach (Zienkiewicz and Taylor, 1977; Altintas et al, 2005). Originally, this method is used for the description of continuous transformation of machined material regarding e.g. its deformation or the change in temperature. Therefore, the finite elements method is particularly suitable for materials which allow plastic or elastic deformation, respectively effects which cause a continuous change on the material. In the case of rigid materials, like natural stone, the material removal is a discontinuous process resulting in brittle fracture and discontinuous chip formation (Denkena et al, 2004). Such situa-

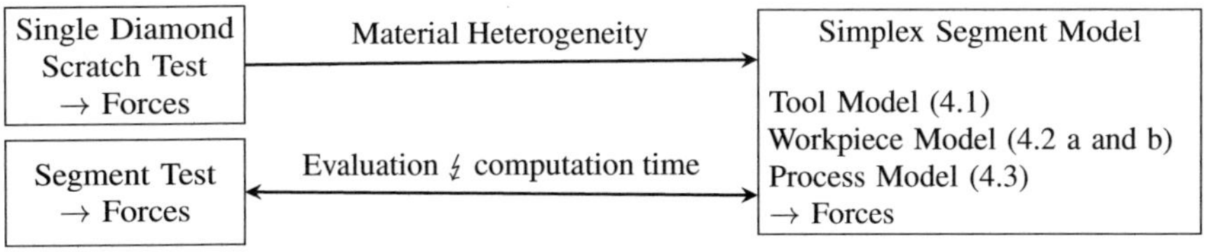

Fig. 1 Development of the Simplex Segment Model.

tions are often solved by the discrete elements approach or by a combination of both methods (Munjiza et al, 1995).

Raabe et al. (2011) considered a model for the force simulation of a single diamond grinding process with a geometrically undefined cutting edge scratching on basalt. Their approach is closely related to the discrete elements method since the removal mechanism is simulated by removing parts from a workpiece represented by a set of 3-dimensional simplexes. The resulting forces are calculated using a geometrical approach involving the angles of the interacting simplexes of diamond and workpiece. In the following papers the model was extended by including the material heterogeneity (Raabe et al, 2012) and compared with experimental data (Weihs et al, 2014). Continuing this work we introduce two models with further extensions concerning the material (from basalt to concrete) and the tool (from single diamond to a segment).

2 Outline

In this work we present two different models (Simplex Segment Model in Sect. 4 and Scratch Track Model in Sect. 5) for the machining of concrete with a single tool segment. The concrete is assumed to consist of two aggregates, basalt and cement, while the segment is a sintered composite of uniformly distributed diamonds in a metal matrix (see Sect. 3 for details).

In Sect. 4 we present a model (Simplex Segment Model, see Fig. 1 for model development) for segment grinding as an extension of the single diamond model of Raabe et al (2011, 2012) and Weihs et al (2014). Two models for the workpiece simulation are described in Sect. 4.2. The representation of the multi-phase tool (segment) is explained in Sect. 4.1, followed by a draft version of the process simulation (Sect. 4.3), where we discuss the computational challenge concerning the simulation time of this segment model.

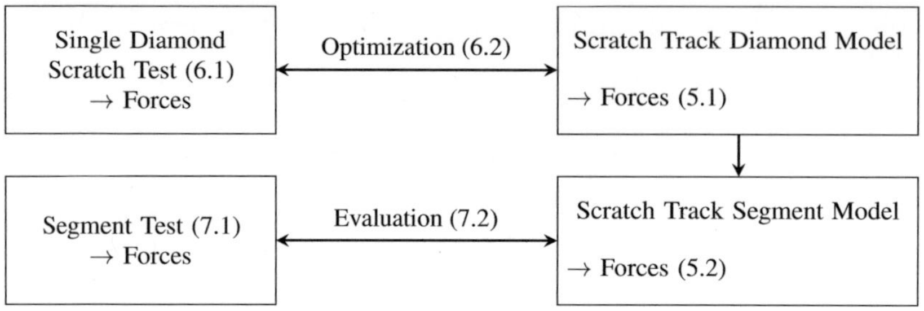

Fig. 2 Development of the Scratch Track Model.

The computation time of this computationally expansive simulation is reduced by introducing an assumption about the geometry of the diamond's cutting profile. While the Simplex Segment Model is working with geometrically undefined cutting edges of the diamonds in the segment, the shape of the diamonds in the Scratch Track Model (see Fig. 2) in Sect. 5 is restricted to pyramids. This assumption allows the modeling of the scratch track which results when one or more diamonds scratch the surface of the workpiece. The development proceeds in two steps. In the first step we adjust all scratch track diamond model parameters by minimizing the deviation between observed forces from single diamond scratch tests and forces of the Scratch Track Diamond Model. For this we will first introduce the scratch track diamond model in Sect. 5.1, describe the experiments (Sect. 6.1) and then explain the optimization procedure and the results in Sect. 6.2. The second step is a feasibility study (Sect. 7.2) comparing the forces of the Scratch Track Segment Model (Sect. 5.2) with the forces of conducted experiments with segments (7.1).

The goal of this work is to predict the arising forces while drilling with the segment into concrete up to a predefined total depth with a constant cutting speed and a constant feed speed.

3 Grinding Process

The core drilling process is a widely used method in the construction industry. For this work diamond tipped drill core bits are used. A drill core bit consists of several rectangular segments attached to a circular body in equally spaced intervals. Each segment is a sintered composite of diamonds and metal powder. Due to a large number of influencing factors, measurement results gained from

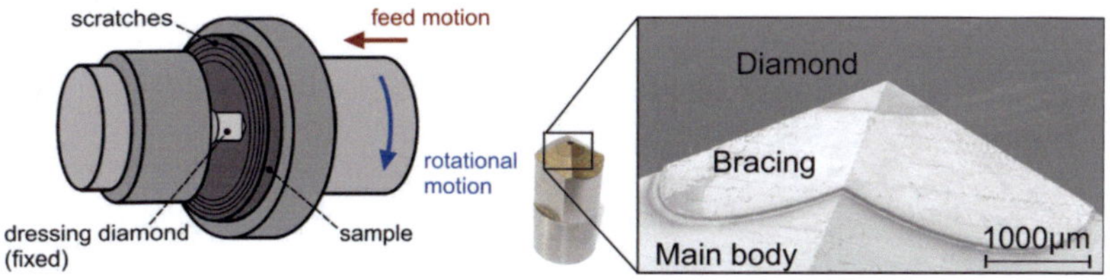

(a) Scratch test (b) Dressing diamond

Fig. 3 Experimental setup for single diamond scratch test.

drilling tests provide only encapsulated information, because dependencies and interfaces cannot be distinguished clearly (Franca et al, 2015). Hence, the first logical step is to reduce the influencing factors by reducing the number of segments which are used for a drilling operation and therefore reducing the number of diamonds. Consequently, two different analysis approaches are studied. Tests with single diamonds, called scratch tests and tests with single segments comprising a number of diamonds on the surface.

Single Diamond Scratch Test To gain a better and more fundamental understanding of the complex grinding process, scratch tests with single diamonds are conducted (Fig. 3 (a)). The advantage of this procedure is the better process control due to the absence of diamond break outs, interactions between diamonds, and the influence of the metal matrix surrounding the diamonds in a segment. In the experimental setup a diamond with a pyramidal shape (Fig. 3 (b)) scratches on a circular path with radius r $[mm]$, a constant cutting speed v_c $\left[\frac{m}{min}\right]$ and a constant feed speed v_f $\left[\frac{mm}{min}\right]$ into the specimens until a total depth is reached. During the experiment the forces (tangential force f_x, radial force f_y, normal force f_z) are recorded.

Single Segment Test For single segment tests, the segments are manufactured in a powder metallurgical process route as a mixture of diamonds and metal powder. During the experiment the segment is attached to a tool holder (Fig. 4), so that the diamonds with workpiece contact scratch the workpiece on a circular path. As in the single diamond tests the cutting and feed speed are constant and the forces (tangential force f_x, radial force f_y, normal force f_z) are recorded.

Material Concrete is a composite material which consists of three main constituents: cement, water and aggregates. Due to chemical reactions between water and cement a hardening process occurs so that the cement acts like a

Fig. 4 Experimental setup for segment test.

binder which holds the aggregates together and builds a strong connection. Cement stone is a mixture of sand and water without aggregates like basalt.

4 Simplex Segment Model

The simplex model is a direct extension of the Raabe et al. (2011, 2012) model and consists of the three parts: tool, workpiece and process simulation.

4.1 Multi-Phase Tool

Diamond Assuming that all diamonds used for the segment production have the shape of truncated octahedra with different edge lengths, the two parameters $l_k = \frac{g}{2\sqrt{2}}$ and $c_k \in (0, l_k]$ (see Fig. 5 (a)) determine the geometrical form of a single diamond with size g. For simplicity in simulation, the truncated octahedron is subdivided into 3-dimensional simplexes as shown in Fig. 5 (c) by applying a Delaunay tessellation (Barber et al, 1996). Simplexes can be used to simulate the diamond wear by removing single simplexes from the diamond's simplex set. When considering the diamond wear, simplexes should be small and numerous. Since size and number of simplexes depend on the number of points used for the tessellation, such points have to be placed inside the truncated octahedron either at random positions or by creating a $3D$-lattice (Fig. 5 (b)). The lattice can be generated, e.g., by stringing together cubic diamond crystal structures as it was proposed by Raabe et al. (2012). The last step in the diamond simulation is a random rotation of all points.

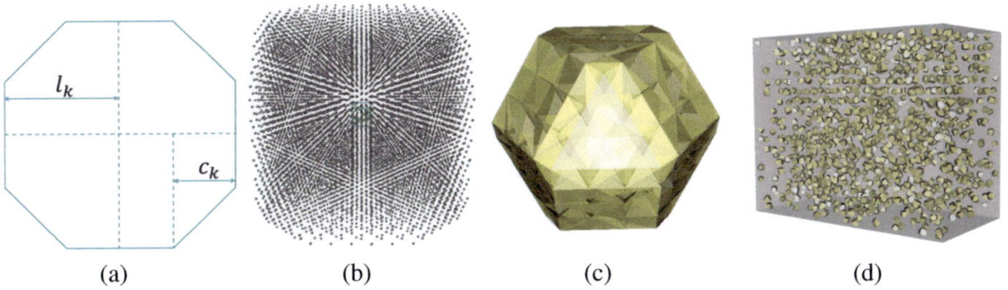

Fig. 5 (a) shape parameters, (b) cubic crystal structure lattice (c) simulated diamond (d) simulated segment with 5 vol.-% diamonds

Segment When moving from a single diamond towards a complete drill core bit, the segment is an intermediate step. As a sintered composite of diamonds and metal powder it introduces new parameters. Design parameters are shape and size of the segment, the size distribution of the diamonds, and their volume fraction ρ in the segment. Suppose, e.g., the diamond sizes g are uniformly distributed between 0.3 and 0.4 mm (equates to 40/50 mesh) and there are 5 vol.-% diamonds in the segment of size $a \times b \times c$ and volume $V_S = abc$. For the expected diamond size $E(g)$ the volume of this diamond is given by $V(E(g)) = 8\sqrt{2} \cdot 10^{-\frac{3}{2}} E(g)^3$. To get a diamond volume fraction of ρ there have to be

$$m = \frac{V_S}{V(E(g))}\rho$$

diamonds of size $E(g)$ in the segment. Therefore, we sample $\lfloor 2m \rfloor$ diamond sizes and determine the corresponding volumes $V(g_1), \ldots, V(g_{\lfloor 2m \rfloor})$. Sampling more sizes than probably needed provides us with the flexibility to reasonably approximate the volume fraction ρ. To achieve this, the first

$$n = \underset{1 \le i \le \lfloor 2m \rfloor}{\arg\min} \frac{\sum_{k=1}^{i} V(g_k)}{V_S} - \rho$$

sizes are taken for the diamonds placed in the segment. The positions $p_1, \ldots, p_n$ for these diamond sizes are sampled under the condition

$$\left\| p_k - p_j \right\| \ge \frac{g_{(k)} + g_{(j)}}{2} \quad \forall j < k$$

to guarantee that the diamonds do not overlap each other. For these positions the diamonds with sizes $g_1, \ldots, g_n$ are simulated as described above. A result is shown in Fig. 5 (d).

4.2 Multi-Phase Workpiece

In this section we will present two ideas for workpiece simulation which allow simulating concrete as a composite of different materials like basalt and cement, and reinforced concrete. The workpiece has the shape of a hollow cylinder with a height h and radii $r \pm b = \frac{d_P}{2} \pm b$, where the half width b of the cylinder must be greater than half the diamond size or half the segment width (Fig. 6 (a)).

If we want to simulate reinforced concrete, we first need to simulate the reinforcing bar with diameter d_S. The position of this bar is given by an axis passing through two predefined or random points. Around this axis a point lattice is expanded. Then we create an equidistant cement grid with point distance δ_{coarse} on

$$[-\lceil r+b \rceil, \lceil r+b \rceil] \times [-\lceil r+b \rceil, \lceil r+b \rceil] \times [0,h]$$

in steel direction to avoid irregular spacing between the bar and cement. To fill the cement grid with basalt grains, we repeat the next two steps until the desired basalt volume fraction is achieved. First we sample a random point from our coarse grid and a random grain diameter from the basalt diameter distribution $U(a_{\text{bas}}, b_{\text{bas}})$, where a_{bas} and b_{bas} are the lower and upper bound of the occurring basalt grain diameters. If there are no other grains (or steel) overlapping the sphere with the defined diameter around this random point, we define all points inside as basalt grain. The resulting workpiece grid is shown in Fig. 6 (b). Due to different material properties and inhomogeneity within the same material each point receives an intrinsic value according to its material. To achieve this, material specific exponential covariance functions are fitted from the estimated seasonality of the force time series of real basalt and cement experiments (Raabe et al, 2012). To use this information for each basalt grain grid and the remaining cement grid Gaussian random fields are sampled with the fitted covariance functions (see Fig. 6 (c)).

Approach a In the first steps all calculations are done on the coarse grid to save time. Since we want to degrade the workpiece into fragments by applying a Delaunay tessellation on the set of points from our grid, the distance between

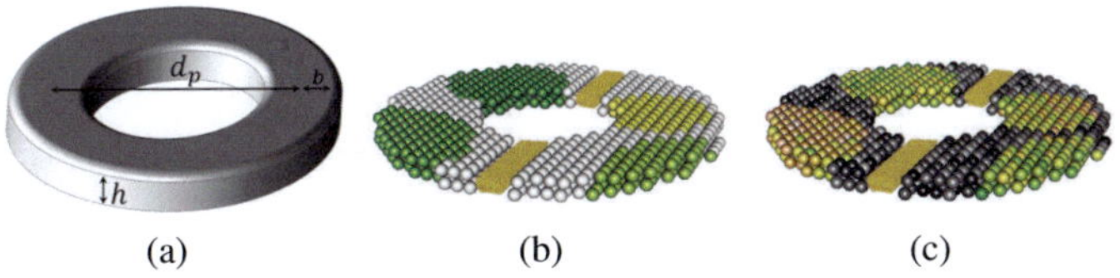

Fig. 6 (a) Basic shape, (b) coarse grid with steel points (yellow), cement points (grey) and basalt points, (c) coarse grid after point elimination and (d) coarse grid with values (represented by different color shades) from sampled Gaussian random fields

Fig. 7 (a) Finer grid with interpolated point values, (b) tessellation of basalt grains and reinforced bar, (c) tessellation of cement points and boundary points of the objects and (d) complete workpiece.

the points influences the size of the resulting tetrahedra. Due to the fact that the chip size (size of the removed material fragments) is very small we need a finer grid with point distance $\delta_{fine} < \delta_{coarse}$. The values for these grid points are interpolated by ordinary Kriging from the values of the coarse grid (Fig. 7 (a)). The Delaunay tessellation of the finer grid proceeds in two steps. We first apply it to the different workpiece objects (basalt grains, steel bar, Fig. 7 (b)). Then the remaining cement points and the boundary points of basalt and steel are degraded into simplexes (Fig. 7 (c)). It is obvious that especially in the second part of the workpiece tessellation the set of points is not convex. To handle this problem we remove all simplexes with maximal edge length greater than the 0.98–quantile of all maximal simplex edge lengths.

This procedure works much better than a tessellation of all points at once because it respects the boundaries of the single objects. Finally, each simplex receives the mean value of its four points' values which are of the same material as the simplex.

Approach b The most time consuming factor in the method of approach (a) is the Delaunay tessellation. To reduce this we provide a different approach. As described above we still need a finer grid but instead of expanding a finer grid over the whole workpiece shape, we just take one part of the hollow cylinder with the correct angle, being a fraction of π. By the Delaunay tessellation of

(a) (b) (c) (d)

Fig. 8 (a) Workpiece blank, (b) first workpiece part with assigned simplex values, (c) four aligned workpiece parts and (d) complete workpiece consisting of sixteen aligned parts.

Table 1 Average values for 100 simulated concrete workpieces of the same size (standard deviation in parentheses).

	Approach a	Approach b
Points in finer grids	7504	11264
Point distance $[mm]$	0.33	0.251
Simplices	47578 (250)	47568 (0)
Mean simplex volume $[mm^3]$	$4.253 \cdot 10^{-3}$ $(9.973 \cdot 10^{-6})$	$2.623 \cdot 10^{-3}$ $(3.304 \cdot 10^{-6})$
Simulation time $[sec]$	37.147 (2.743)	10.716 (0.321)

this grid part we receive a degraded workpiece sector as shown in Fig. 8 (a) without point or simplex values. Since neither the points nor the simplexes of this sector have assigned values, we will call it a 'blank'. To create the workpiece, the next three steps have to be repeated until the hollow cylinder is complete (Fig. 8 (d)). A copy of the blank with jittered points is rotated to its position in the workpiece. Then we interpolate the values for the points of this part by ordinary Kriging from the values of the random fields of the coarse grid. Here we use the information about positions and sizes of the basalt grains for a material separated interpolation. Afterwards the values for the simplexes are calculated from the point values (Fig. 8 (b)). Because of using copies of the one blank, each part of the workpiece has the same Delaunay tessellation. Nevertheless, all simplexes have different volumes because we changed the basis of the tessellation by jittering the points in each part.

To compare the two workpiece simulations one hundred concrete workpieces with the sizes $d_p = 20\ mm$, $b = 2.5\ mm$ and a height of $h = 1\ mm$ were simulated for both procedures. Despite the fact that the numbers of simplexes are rather similar (see Table 1), the simulation time required for the second procedure is much shorter, as intended. Another advantage of the second workpiece simulation is that we have no variation in the number of simplexes because the blank tessellation does not depend on the material. That makes it easier to calculate the needed memory size.

4.3 Process Simulation

To simulate the machining process with workpieces as in Sect. 4.2 we first have to simulate a workpiece of desired size and material and a diamond or segment. After positioning the diamond or segment on the surface of the workpiece the process starts with the first movement of the tool. The length and depth of this movement depends on the cutting speed v_c, cutting depth per revolution $a_p = 10^{-3}\frac{v_f}{v_c}2\pi r\ [mm]$ and the number of iterations ν per revolution. For the simulation of N revolutions, we have to determine for each of the νN iterations whether the simulated tool has contact with the simulated workpiece. In this case the affecting forces are computed. When using a tool segment machining concrete, there are four possible interactions: basalt-diamond, basalt-metal matrix, cement-diamond and cement-metal matrix.

For the force calculation we can use a geometrical approach based on the edge orientation of the colliding simplexes and the division of the resulting force into radial and normal force described for the process with a single diamond in Raabe et al. (2012). In this approach each workpiece simplex hit by a diamond simplex is removed from the simulated workpiece. To extend this procedure to the grinding with a segment, we assume that material removal is only caused by the diamonds and not by the metal matrix. With the additional assumption that the force time signal is dominated by the forces arising in diamond-workpiece interaction, we only have to distinguish between the different workpiece materials. Nevertheless, in each iteration we have to determine each workpiece simplex with non-zero intersection volume with a simplex of at least one of the diamonds in the segment. The computation time of one iteration step depends on the number of simplexes in all diamonds and the number of simplexes in the workpiece. Since the whole number of simplexes decreases due to the wear and removal simulation, the evaluation of later iterations is faster. At the end of the process simulation many workpiece simplexes outside the scratch track will remain because they were not hit by any of the diamonds.

Unfortunately, it turned out that this model only appears to be appropriate for the simulation of short single diamond experiments but not for the much more complex simulation of segment experiments, which require the simulation of hundreds of revolutions with multiple diamonds. For this purpose, we developed another approach.

5 Scratch Track Model

In the new approach we reduce workpiece modeling to a minimum. Instead of modeling the complete workpiece and then remove parts of it, we only simulate the parts which are removed by the diamonds.

The shape of the diamond, introduced in Sect. 4.1, is simplified to a pyramid turned upside down as used in the single diamond experiments (Sect. 3). This simplification to a pyramidal form can be justified since the part of the octahedral diamond form that removes material in the segment experiments is very similar to a rotated pyramid. By this assumption, the resulting scratch track has the profile of a triangle with angle α determined by the cutting profile of the diamond.

Before we introduce the force model for the grinding process with a segment (5.2), we explain the model idea for the special case of a single diamond (5.1).

5.1 Scratch Track Diamond Model

In the one diamond case, a single diamond is scratching on a circular path along the workpiece surface (see Sect. 3). The maximal intrusion depth of the diamond is limited by the height of the diamond. For simplification we assume that this maximal depth is reached after N revolutions. Let denote v the number of modeled observations per revolution and $a_p = \frac{v_f}{v_c} 2\pi r$ the cutting depth per revolution. It is obvious that the resulting forces depend at least on the volume of removed material and characteristics of the machined material. Thus, in our model a realization of the modeled force is obtained by

$$F_i = \frac{g_{zv}}{r} \cdot z_i \cdot v_i + \frac{g_v}{r} \cdot v_i, \quad i = 1, \ldots, Nv,$$

where r is the drilling radius, g_{zv} and g_v parameters, which have to be optimized, v_i the volume removed from the workpiece and z_i the material heterogeneity (Herbrandt et al, 2016). In the following we will describe the concept of the scratch track model and how the information about removed volume and material characteristics is linked to this scratch track.

To simulate v observations per revolution and N revolutions in total for one diamond, we place $vN+1$ triangles evenly distributed along the diamond's scratch track (Fig. 9 (a)), whereby the first triangle has an area of zero. The sizes of these triangles depend on the intrusion depth of the diamond in the workpiece and the angle α of the pyramid representing the shape of the diamond. Since the cutting depth per revolution a_p is known, we can assume that the intrusion depth increases by $a = \frac{a_p}{\nu}$ for each simulated scratch track triangle D_j $(j = 1, \ldots, \nu N + 1)$ (see Fig. 9 (a) with the triangles $D_1, \ldots, D_{21}$ of the first revolution with $\nu = 20$). To consider the brittleness of the material, we allow a $Beta(0, a_p, p, q)$-distributed size variation $a^\star$ of the triangles, where $Beta(0, a_p, p, q)$ is the generalized beta distribution for the interval $[0, a_p]$ with the unknown parameters p and q. Thus, the height of the jth triangle D_j with the corner points (d_{j1}, d_{j2}, d_{j3}) is $h_j = a(j-2) + a_j^\star$ $(j = 2, \ldots, N\nu + 1$ and $h_1 = 0)$, where the i.i.d. variables $a_2^\star, \ldots, a_{\nu N+1}^\star$ have the same distribution as $a^\star$.

A simulated observation is represented by a scratch track part formed by the connection of two adjacent triangles (see Fig. 9 (b)). The connection is realized by three 3-dimensional simplexes and the removed volume v_i is calculated as the sum of the volumes of the three simplexes.

As in Sect. 4.2 the material heterogeneity is considered by sampling from Gaussian random fields. In contrast to Sect. 4.2, however, the number of values we have to sample from the Gaussian random fields is smaller, since only the $3(N\nu + 1)$ points of the $N\nu + 1$ triangles are taken into account. Additionally, we want to adjust the parameters $\mu, \sigma^2, \sigma_\xi^2, \psi$ of the Gaussian random field together with all the other parameters (p, q of the Beta distribution and g_{vz}, g_v) by minimizing the deviation between the observed and modeled forces (see Sect. 6.2). The material heterogeneity z_i of the ith modeled observation is calculated as the mean of the six sampled point values of each two adjoining triangles (d_{j1}, d_{j2}, d_{j3}) and $\left(d_{(j+1)1}, d_{(j+1)2}, d_{(j+1)3}\right)$. In Fig. 9 (c) the six values are represented by the colors of the six points of the two triangles and the overall mean is represented by the color of the polyhedron (scratch track part) resulting by the connection of these two triangles. Fig. 9 (d) shows 100 of thus scratch track parts with heterogeneity values represented by colors in the first revolution.

5.2 Scratch Track Segment Model

In a segment we have several diamonds at random positions (see Sect. 3 and Sect. 4.1). In addition to the first assumption (diamond shape), we introduce a further assumption concerning the scratching with more than one diamond at

(a) (b)
(c) (d)

Fig. 9 (a) Scratch track triangles for the first revolution and (b) tessellation of one scratch track part into three simplexes (red, green and blue), (c) Scratch track part with sampled heterogeneity values for the six points represented by different color shades for the points and the mean value (color of track part) and (d) scratch track of the first revolution with $\nu = 100$ observations (scratch track parts) with assigned heterogeneity values.

the same time. The second assumption states that the scratch tracks of different diamonds are independent of each other.

For our model one of the most important differences between grinding with a diamond and grinding with a segment (which corresponds to grinding with several diamonds) is the maximal intrusion depth. In the single diamond case this depth is determined by the diamond height, since the one diamond defines the complete tool. Therefore the intrusion depth ranges from 0 to the height of the diamond $h_D = \frac{g}{2\tan\frac{\alpha}{2}}$, where g is the diamond size and α the pyramid angle. In the segment experiment the diamonds are held by the metal matrix. Suppose that the position of the lowest diamond of size g and height h_D is $p = (p_x, p_y, p_z)^T$ (Fig. 10 (a)). Then the first contact of this diamond with the workpiece is at the intrusion depth of $p_z - h_D$ (Fig. 10 (b)). The Figs. 10 (b)–(d) show the intrusion period of this diamond. The diamond of height h_D is completely in the workpiece at the cutting depth of p_z (Fig. 10 (d)). When this maximal intrusion depth of the diamond is reached, the diamond is still held by the metal matrix and the grinding process proceeds (Fig. 10 (e)). We assume that the diamond breaks out at an unknown cutting depth p_z'.

The scratch track diamond model needs some adaptations for the segment application. The adaptations of the three model parts scratch track, volume and heterogeneity will be discussed in the same order as in Sect. 5.1.

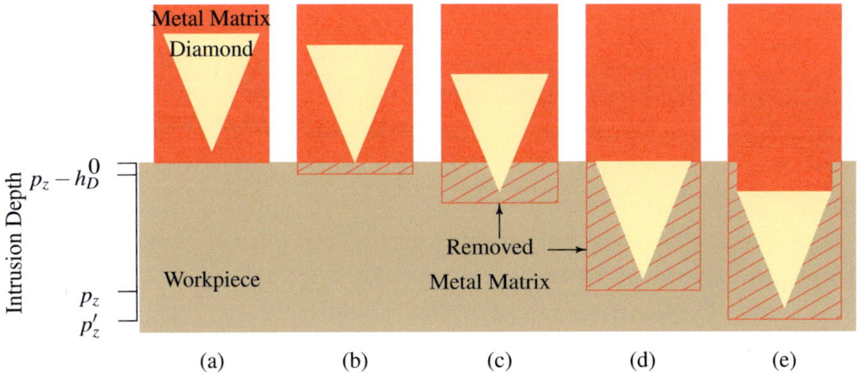

Fig. 10 (a) First segment workpiece contact with a diamond (yellow) inside the metal matrix (red), (b) First diamond workpiece contact at a cutting depth of $p_z - h_D$, (c) Diamond in the intrusion period, (d) Diamond at the end of the intrusion period at a cutting depth of p_z, (e) Last position of the diamond before break out at p'_z.

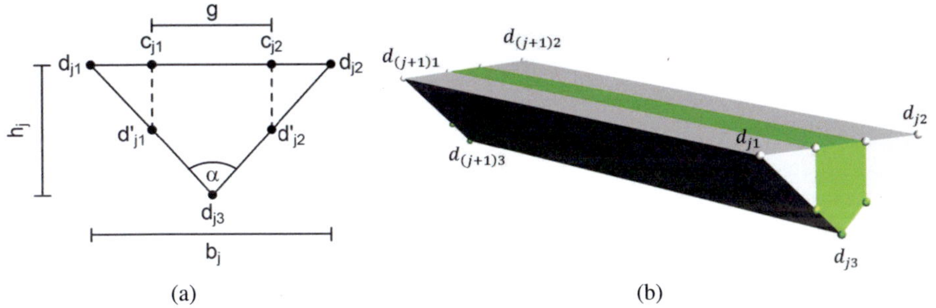

Fig. 11 (a) jth scratch track triangle $D_j = (d_{j1}, d_{j2}, d_{j3})$ and diamond profile $\left(d'_{j1}, d'_{j2}, d_{j3}\right)$ and (b) ith scratch track part

When modeling the scratch track we have to consider the case in Fig. 10 (e). When the length of the triangle's base $b_j = 2h_j \tan\frac{\alpha}{2}$ (h_j height of the jth triangle) representing the part of the diamond inside the workpiece exceeds the size of the diamond g, the intrusion period of the diamond has ended (as shown in Fig. 10 (d)). Since the diamond profile will not increase any more, we have to cut off the corners $\left(d_{j1}, d'_{j1}, c_{j1}\right)$ and $\left(d_{j2}, d'_{j2}, c_{j2}\right)$ (see Fig. 11 (a)) of the following scratch track triangles. The diamond's profile in the workpiece is determined by the triangles (d_{j1}, d_{j2}, d_{j3}) during the intrusion period and then by $\left(d'_{j1}, d'_{j2}, d_{j3}\right)$.

After the intrusion period the volume is reduced by the volumes v_{i1} (volume of the right grey polyhedron in Fig. 11 (b)) and v_{i2} (volume of the left grey polyhedron in Fig. 11 (b)) between the corresponding cut off corners. The resulting volume is the volume of the green polyhedron in Fig. 11 (b). Additionally, the volume v_i is reduced by the already removed volume $v_{i-\nu}$ in the previous revolution for $i > \nu$ (same procedure as for the scratch track diamond model).

After that, the material heterogeneity z_i of the i-th modeled observation is calculated as the mean of the six sampled point values of each two adjoining diamonds' profile triangles (d_{j1}, d_{j2}, d_{j3}) and $(d_{(j+1)1}, d_{(j+1)2}, d_{(j+1)3})$ (as in the scratch track diamond model) or $(d'_{j1}, d'_{j2}, d_{j3})$ and $(d'_{(j+1)1}, d'_{(j+1)2}, d_{(j+1)3})$ after the intrusion period of the diamond (Fig. 11 (b)).

The resulting normal forces

$$F_i = \begin{cases} \frac{g_{zv}}{r} \cdot z_i \cdot v_i + \frac{g_v}{r} \cdot v_i, & p_z - h_D \leq a_{i+1} \leq p'_z \\ 0, & \text{otherwise} \end{cases}$$

are modeled so that for one diamond scratching at radius r and parameters g_{zv} and g_v normal forces increase as the removed volume v_i increases, while the variance is represented by the heterogeneity values z_i ($i = 1, \ldots, N\nu$). For K diamonds the total force in the ith iteration

$$F_{i,\text{total}} = \sum_{k=1}^{K} F_{i,k}$$

is determined as the sum of the K forces $F_{i,1}, \ldots, F_{i,K}$ in the ith iteration.

For the simulation of $N\nu = 4500$ force observations with a segment including one diamond by using the presented scratch track segment model we need 9.83 (± 0.997) seconds. In almost the same time we can simulate the workpiece of the simplex segment model (approx. 10 seconds, see Table 1 in Sect. 4.2). For the simulation of 4500 observations with the simplex segment model we additionally need to simulate the process (Sect. 4.3) to calculate the forces. That means the scratch track segment model has finished the computation of 4500 observations even before the simplex segment model is ready to compute the first observation.

6 Single Diamond Grinding

In this section we will focus on the Scratch Track Diamond Model (Sect. 5.1). For the model parameter adjustment we first explain the details of the conducted single diamond experiments (Sect. 6.1) which provide the force data we use as reference in the adjustment procedure. Then the optimization of the model parameters and the results are presented (Sect. 6.2).

6.1 Design of Single Diamond Experiments

In the experimental setup a diamond scratches into the specimens until a total depth of $A = 0.08\ mm$. During the experiment the forces (tangential force f_x, radial force f_y, normal force f_z) are recorded with a sampling rate of $v_f = 200000\ Hz$. Depending on the total drilling depth and the speeds v_c and v_f, the total number of recorded observations per experiment and force can be calculated as $\frac{A \cdot 60}{v_f} v_f$ (here: between 101000 and 480000). The tests are conducted on a machining center (IXION TLF 1004) without a coolant or lubricant. For the analysis of the influence of the cutting speed $v_c \left[\frac{m}{min}\right]$ and the feed speed $v_f \left[\frac{mm}{min}\right]$ on the resulting process forces, a 4^2- full factorial design with the parameter setting $v_c \in \{40.5, 117, 193.5, 270\}$ and $v_f \in \{2, 4.5, 7, 9.5\}$ is chosen. By carrying out scratch tests on single phases of the composite material concrete the process is subdivided into subprocesses. Hence, tests on single phase basalt and cement stone are conducted to analyze the forces developing during the scratching. Five samples of each material are available and each of them can be scratched on 12 radii $r \in \{16, 17, \ldots, 27\}\ mm$. The destructive testing does not allow real repetitions, so each speed combination (v_c, v_f) is repeated on adjacent radii of a sample. The 16 speed combinations of the full factorial design are distributed to six blocks of size five using the $D-$criterion.

Let denote $\mathscr{R}(v_c, v_f)$ the set of radii with the same speed combination (v_c, v_f) and $n_{\mathscr{R}}(v_c, v_f)$ the number of elements in this set. Since each speed combination is repeated on the adjacent radius, each set contains at least two elements.

6.2 Optimization of the Scratch Track Diamond Model

To find out whether the approach in Sect. 5 is suitable to describe forces arising during a single grain scratch test, the model parameters $\theta = (g_{zv},\ g_v,\ \mu,\ \sigma^2,\ \sigma_\xi^2,\ \psi,\ p,\ q)$, where $\mu, \sigma^2, \sigma_\xi^2, \psi$ are the parameters of the Gaussian random field, are adjusted to the normal forces ($f_z = f$) from the conducted single grain experiments (see sec. 6.1) on basalt and cement (see Herbrandt et al, 2016, for more details). The adjustment is performed for each speed combination (v_c, v_f) by applying model based optimization techniques which are particularly suitable for the optimization of expensive black box functions (Jones et al, 1998). The target is the minimization of the objective function determining the deviation between observed and modeled forces. For this purpose the expected deviation

$$E\left(\left\|f\left(v_c, v_f, r\right) - \mathscr{F}\left(\theta, r\right)\right\|_D\right) = E\left(D\left(f\left(v_c, v_f, r\right), \mathscr{F}\left(\theta, r\right)\right)\right) \qquad (1)$$

of a measured force $f\left(v_c, v_f, r\right)$ from the model force $\mathscr{F}\left(\theta, r\right)$ (underlying force model process with realizations $F\left(\theta, r\right)$ as described in Sect. 5) is minimized. By estimating the expectation with the arithmetic average of M (here: $M = 25$) realizations F of the force model $\mathscr{F}$ and $n_{\mathscr{R}}\left(v_c, v_f\right)$ observed forces $f\left(v_c, v_f, r\right)$ with radii $r \in \mathscr{R}\left(v_c, v_f\right)$, the optimal parameter settings for one speed combination are obtained as

$$\begin{aligned} \theta^\star\left(v_c, v_f\right) &= \underset{\theta \in \Theta}{\arg\min}\ \bar{D}\left(f\left(v_c, v_f\right), F\left(\theta\right)\right) \\ &= \underset{\theta \in \Theta}{\arg\min} \frac{1}{3Mn_{\mathscr{R}}\left(v_c, v_f\right)} \sum_{r \in \mathscr{R}\left(v_c, v_f\right)} \sum_{m=1}^{M} \left[d_R\left(\widetilde{f}\left(v_c, v_f, r\right), \widetilde{F}\left(\theta, r\right)\right)\right. \qquad (2) \\ &\left. + d_\beta\left(f^\star\left(v_c, v_f, r\right), F^\star\left(\theta, r\right)\right) + d_S\left(\widetilde{f}\left(v_c, v_f, r\right), \widetilde{F}\left(\theta, r\right)\right)\right], \end{aligned}$$

where the terms are discussed in the following. The considered deviation measure $\bar{D}$ is the mean of measures for the comparison of the three characteristics slope, range, and spectrum. For the comparison the forces $f = f\left(v_c, v_f, r\right) = \left\{f_{t_i}\left(v_c, v_f, r\right) \mid 0 \le t_i \le T_f,\ i = 1, \ldots, L,\ L \text{ number of observations},\ T_f \text{ observation time in seconds}\right\}$ and $F = F\left(\theta, r\right) = \left\{F_{t_i}\left(\theta, r\right) \mid 0 \le t_i \le T_F,\ i = 1, \ldots, N\nu\right\}$ have to be aligned. Due to the different sampling rates and since the sampling rate of f is very high, we decide to exploit the characteristics of the time series, rather than applying very time consuming methods like the dynamic time warping. Therefore, the forces f and F are aligned by the intercepts of the corresponding linear models

$$f = \beta_{f0} + \beta_{f1}t + \varepsilon_f \text{ and } F = \beta_{F0} + \beta_{F1}t + \varepsilon_F. \tag{3}$$

Therefore, the force with the smaller estimated intercept ($\widehat{\beta}_{f0}$ or $\widehat{\beta}_{F0}$) is shifted by redefining the starting time

$$f^{\star} = \begin{cases} f, & \widehat{\beta}_{f0} > \widehat{\beta}_{F0} \\ \left\{ f_{t_i} \mid 0 \leq t_i - \frac{\widehat{\beta}_{F0} - \widehat{\beta}_{f0}}{\widehat{\beta}_{f1}} \leq T_f - \frac{\widehat{\beta}_{F0} - \widehat{\beta}_{f0}}{\widehat{\beta}_{f1}} = T_{f^{\star}} \right\}, & \widehat{\beta}_{f0} \leq \widehat{\beta}_{F0} \end{cases} \tag{4}$$

(F analogue). For the comparison of range and spectrum the forces are additionally detrended, so that

$$\widetilde{f} = \left\{ f^{\star}_{t_i} - \widehat{\beta}_{f^{\star}0} - \widehat{\beta}_{f^{\star}1} t_i \mid 0 \leq t_i \leq \min\left\{ T_{f^{\star}}, T_{F^{\star}} \right\} = T_{\widetilde{f}} \right\} \tag{5}$$

with $f^{\star} = \beta_{f^{\star}0} + \beta_{f^{\star}1}t + \varepsilon_{f^{\star}}$ (F analogue). Then the range difference is

$$d_R\left(\widetilde{f}, \widetilde{F}\right) = \left| \max_{0 \leq t \leq T_{\widetilde{f}}} \widetilde{f}_t - \min_{0 \leq t \leq T_{\widetilde{f}}} \widetilde{f}_t - \max_{0 \leq t \leq T_{\widetilde{F}}} \widetilde{F}_t + \min_{0 \leq t \leq T_{\widetilde{F}}} \widetilde{F}_t \right| \tag{6}$$

and the slope difference is

$$d_{\beta}\left(f^{\star}, F^{\star}\right) = \left| \widehat{\beta}_{f^{\star}1} - \widehat{\beta}_{F^{\star}1} \right|. \tag{7}$$

Since the modelled sampling rate

$$\nu_F = \frac{v v_c 10^3}{2\pi r 60} \quad \left[\frac{1}{s}\right] \tag{8}$$

is much smaller than the sampling rate ν_f $\left[\frac{1}{s}\right]$ and we consider both time series on the same time interval, the number of considered observations $n_{\widetilde{F}}$ of F is also smaller than $n_{\widetilde{f}}$. Therefore, the spectral differences are only calculated at the Fourier frequencies

$$\varphi_j = \frac{j}{n} \text{ with } n = n_{\widetilde{F}} + \left| \min_{(a,b,c) \in \mathbb{N}^3} n_{\widetilde{F}} - 2^a 3^b 5^c \right| \text{ and } j = 1, \ldots, \left\lfloor \frac{n}{2} \right\rfloor \tag{9}$$

of the shorter time series $\widetilde{F}$. This approach allows the application of the fast Fourier transform (Bloomfield, 2004) algorithm, which by itself enables a fast computation of the periodogram

$$I_{\widetilde{F}}(\varphi_j) = \frac{1}{\nu_F n_{\widetilde{F}}} \left| \sum_{k=1}^{n_{\widetilde{F}}} \widetilde{F}_k \exp(-i2\pi\varphi_j k) \right|^2 \tag{10}$$

as an estimate of the spectrum of $\widetilde{F}$. By adjusting the angular frequencies $2\pi\varphi_j$ to the sampling rate of the measured signal $\widetilde{f}$, we obtain the periodogram

$$I_{\widetilde{f}}(\varphi_j) = \frac{1}{\nu_f n_{\widetilde{f}}} \left| \sum_{k=1}^{n_{\widetilde{f}}} \widetilde{f}_k \exp\left(-i2\pi\varphi_j \frac{\nu_F}{\nu_f} k\right) \right|^2 \tag{11}$$

of $\widetilde{f}$ at the same frequencies φ_j and can determine the spectral differences

$$d_S\left(\widetilde{f}, \widetilde{F}\right) = \sum_{j=1}^{\lfloor \frac{n}{2} \rfloor} \left| I_{\widetilde{f}}(\varphi_j) - I_{\widetilde{F}}(\varphi_j) \right|. \tag{12}$$

Since the scratch track diamond model is stochastic, the noisy Kriging model is chosen as surrogate in the model based optimization process for the CPU-intensive deviation function (Picheny et al, 2013). A new point for evaluation is proposed by maximizing the augmented expected improvement (Huang et al, 2006). We evaluated 800 parameter constellations θ for each speed combination (v_c, v_f). The first 80 points (initial design) for evaluation of the 800 in total were sampled from a random Latin hypercube.

The results achieved with this method show a good agreement between observed and modeled normal forces. Fig. 12 displays exemplarily the normal force from the conducted experiment for the speed combination $\left(v_c = 270\ \frac{m}{min}, v_f = 7\ \frac{mm}{min}\right)$ and 50 modeled force time series with optimized model parameters. As the figure implies, the modeled forces match the slope and variance of the observed force quite well.

Table 2 shows the optimization results for each of the 16 speed combinations (v_c, v_f). In the most cases the best parameters were found in the first 500 optimization steps. For the speed combinations with higher minimal deviation measures $\overline{D}_{\min}$ we observed discrepancies in the course of the corresponding force time series (repetitions with the same speed combination but on different radii or on different material samples). The described deviation measure results in small values if all 25 realizations of the scratch track diamond model fits in terms of slope, range and spectrum to all observed forces with the same speed combination. If the observed forces with the same speed combination are quite different, the optimal parameters found are a compromise which ensure the best fit of the modeled force for all these observations in terms of average.

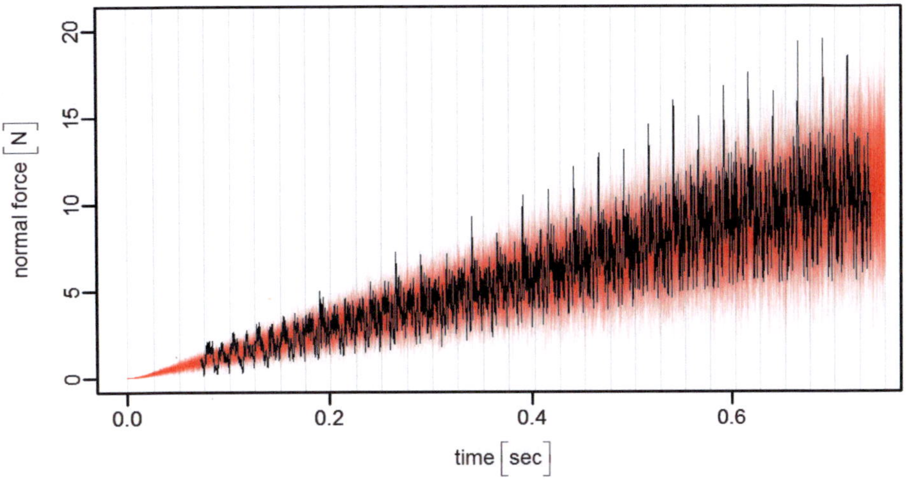

Fig. 12 Normal force from conducted single grain experiment with the parameter settings $v_c = 270\ \frac{m}{min}$, $v_f = 7\ \frac{mm}{min}$, $r = 18\ mm$ (black) and 50 modeled forces (red).

Fig. 13 is an example of an optimization course for the speed combination $v_c = 270\ \frac{m}{min}$, $v_f = 7\ \frac{mm}{min}$. The first 80 points are the realizations of the deviation measure for the model parameter combinations of the initial design (see upper figure in 13). By using the space filling random Latin hypercube design a good parameter combination with $\bar{D} \approx 2.1$ could already be found within these first 80 points (lower figure in 13). The iterative optimization improved this value in the following up to $\bar{D} \approx 1.45$ after the evaluation of 592 further points.

7 Single Segment Grinding

Since the resulting forces of the scratch track diamond model seems rather promising, we start the analysis of the scratch track segment model with a first feasibility study. In the first Subsect. (7.1) we summarize the technological details concerning the segment manufacturing, as well as the design of experiments for the conducted tests with the fabricated segments. The second Subsect. (7.2) will deal with the feasibility study and its results.

Table 2 Optimization results for the 16 speed combinations (v_c, v_f) with minimum value of $\overline{D}$ found and the according iteration $n_{\min}$ of the optimization procedure.

v_c	v_f	$\overline{D}_{\min}$	$n_{\min}$
40.5	2.0	3.245	328
40.5	4.5	3.037	621
40.5	7.0	5.643	682
40.5	9.5	4.317	299
117.0	2.0	1.492	800
117.0	4.5	3.811	490
117.0	7.0	7.075	427
117.0	9.5	4.413	434
193.5	2.0	1.657	731
193.5	4.5	3.729	171
193.5	7.0	1.749	452
193.5	9.5	6.950	423
270.0	2.0	2.089	122
270.0	4.5	4.750	755
270.0	7.0	1.450	672
270.0	9.5	7.607	261

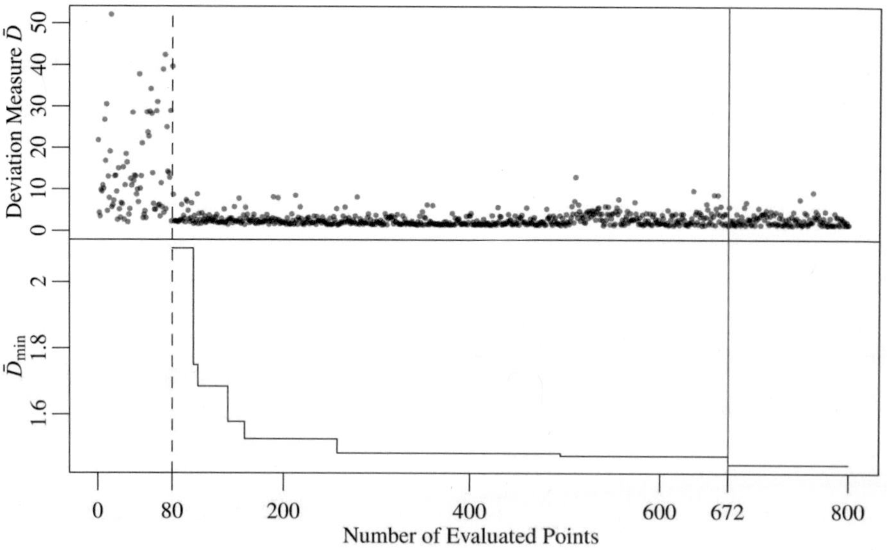

Fig. 13 Optimization course for the speed combination $v_c = 270\ \frac{m}{min}$, $v_f = 7\ \frac{mm}{min}$. Upper Figure: Deviation measure $\bar{D}$ for the 800 parameter combinations of θ. Lower Figure: Minimal deviation measure from the first 80 evaluations (initial design) to all 800 evaluations. Red line marks the evaluation with the best found parameter combinations of θ.

7.1 Design of Single Segment Experiments

In the powder metallurgical process route a four component metal powder consisting of iron, cobalt, copper and tin (Diabase V21, Dr. Fritsch) is used, which is optimized for concrete machining. Within this process route synthetic diamonds (Syngrit SDB1055, Element Six) with varying grain sizes (20/30, 40/50 and 70/80 mesh) are added to the metal powder mixture at variable amounts of 2, 5 and 10 vol.-%. Subsequently the prepared powder-diamond material is homogenized in a tumbling mixer. Finally the raw material is filled in graphite moulds and sintered in a CSP100 hot-pressing facility (Dr. Fritsch) to shape geometries of $8 \times 10\ mm$ rectangles. The maximum pressure is $350 \frac{kg}{cm^2}$ and the sintering parameters are 840° C for three minutes. The single segment tests are carried out on a machining center (FZ 12 S, Chiron) under constant water supply. An additive within the water prevents corrosion of the machining center (Bechem Avantin 361, concentration 7 %). Before testing, segment dressing is carried out in order to expose the first diamond layer of the segment and thus, guarantee the contact between at least one diamond in the segment and the workpiece. The radius of the tool holder amounts to $r = 50\ mm$ (Fig. 4). Force measurements (tangential force f_x, radial force f_y, normal force f_z) are conducted using a force dynamometer (Kistler instruments, type 9255C) with a frequency of $v_f = 10000\ Hz$ until a total depth of $A = 3000\ mm$ is reached. For each diamond grain size and diamond concentration experiments with the parameter settings of a 3^2- full factorial design in circumferential speed $n = \frac{v_c 10^3}{2\pi r} \in \{117, 449, 781\}\ \frac{1}{min}$ (rounds per minute) and feed velocity $v_f \in \{0.5, 1.25, 2\}\ \frac{mm}{min}$ are performed (including repetitions).

7.2 Feasibility Study for the Scratch Track Segment Model

Since the model results for the scratch track diamond model are satisfactory, a feasibility study is established whether the scratch track segment model is able to reproduce a force time series from a conducted segment experiment (see Sect. 7.1). For this purpose the number of active diamonds and the number of broken out diamonds in the segment are fixed to be 15 and 1, respectively, for an experiment with circumferential speed $n = 449\ min^{-1}$, $v_f = 2\ \frac{mm}{min}$, grain size of 40/50 mesh and 2 vol.-% diamond concentration. Additionally, the $x-$ and

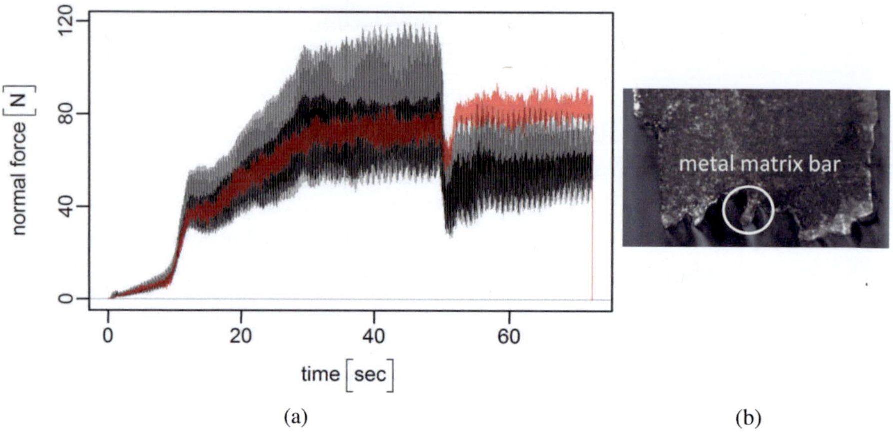

Fig. 14 (a) Normal forces of a segment experiment (black) and modeled forces with optimized parameters (red), (b) segment profile with marked metal matrix bar

$y-$coordinates of the diamonds' positions $p = (p_x, p_y, p_z)^T$ in the segment are measured.

According to the speed parameters the total cutting depth amounts to approximately 2.4 *mm*. Since 14 out of 15 cutting diamonds are still in the segment when the experiment is over, we have to set the end of the diamond-workpiece-interaction p_z' of these 14 diamonds to a value greater than 2.4 *mm*. For the broken out diamond the end of interaction is determined from the structural change in the force time series at approx. 1.67 *mm*. The remaining unknown parameters are the start of interaction of the 14 diamonds (one diamond's interaction starts at 0 *mm*), the grain sizes which are limited to the interval $[0.297, 0.4]$ *mm* corresponding to 40/50 mesh, and the diamonds' profile angles. The adjustment of all these parameters with the model based optimization (as described above) leads to the result presented in Fig. 14 (a). The average course is already well matched up to the point that one diamond breaks out after approx. 50 seconds. There are at least two explanations for this mismatch. One is that the optimized parameter settings are not correct for the broken out diamond. If the chosen grain size is too small or the diamond-workpiece-interaction starts too late, the resulting force of this diamond is too small at the break out point and thus would lead to a too small decrease in the force time series. Another explanation can be referred to a phenomenon that can be observed in segment experiments but not when using a drill core bit, where the material removal and the segment wear are more regular since more segments lead to more cutting diamonds. If a diamond in a single segment is cutting along its circular path at a constant radius and there are no other diamonds at directly adjacent radii, the material to the left and to the right of this diamond is not removed. The result is that the metal matrix of the segment is removed at both sides of the diamond's position and the remaining metal matrix, which is holding the diamond, forms a bar (see Fig. 14 (b)). It is conceivable that the friction between this metal matrix bar and the workpiece results in higher forces. At the break out point the diamond and the metal matrix bar break out which would explain the much smaller forces after the break out. The magnitude of the normal force after the break out of the diamond depends on the size of the diamonds newly active afterwards.

8 Conclusion and Future Work

We have presented two different ways for the simulation of a grinding process. The first approach (in Sect. 4) is based on the tessellation of the workpiece into simplexes and turns out to require too much computation time. A reduction of the workpiece simulation time can be achieved using a slightly different approach (the 'blank'-approach) regarding the tessellation procedure. However, the computation of the process part cannot be accelerated without a substantial loss of accuracy. Therefore, the approach in Sect. 4 is certainly appropriate for the simulation of short single diamond experiments but not for the simulation of segment experiments, which require the simulation of hundreds of revolutions with multiple diamonds.

For this reason the approach in Sect. 5.1 was developed by introducing some assumptions about the scratch track produced by a pyramidal shaped diamond. The model parameters have been successfully adjusted to the data provided by the conducted single grain experiments (Sect. 6.1). Therefore, we performed a feasibility study (7.2) for the segment grinding simulation using the approach in Sect. 5.2 with one of the segment experiments as reference (Sect. 7.1). The average course of the normal force is already well matched by the modeled force. Thus, the presented model can be used as base for further developments. Improvements may be possible regarding the simulation time. Since all scratch track parts are subdivided into three simplexes in the same way, it is possible to derive a closed expression for the volume of each scratch track part and the volume for the scratch track reduction, respectively. This closed expression will make the subdivision into simplexes redundant and thus leads to a faster calculation. We already derived such formulas for the scratch track diamond model (Herbrandt et al, 2016) and we want to extend these results for the presented scratch track diamond model.

Beside the improvement of the actual model, future work will deal with the duration of the diamond-workpiece interaction. Based on the proposed simulations, future experiments will focus on a better understanding of diamond break outs depending on different compositions of the metal powder used for the segment manufacture.

Acknowledgements This work has been supported by the Collaborative Research Center "Statistical Modeling of Nonlinear Dynamic Processes" (SFB 823) of the German Research Foundation (DFG).

References

Altintas Y, Brecher C, Weck M, Witt S (2005) Virtual machine tool. CIRP Annals-Manufacturing Technology 54(2):115–138, DOI 10.1016/s0007-8506(07)60022-5

Barber CB, Dobkin DP, Huhdanpaa H (1996) The quickhull algorithm for convex hulls. ACM Transactions on Mathematical Software (TOMS) 22(4):469–483, DOI 10.1145/235815.235821

Bloomfield P (2004) Fourier Analysis of Time Series: An Introduction. John Wiley & Sons, Hoboken, USA, DOI 10.1002/0471722235

Brinksmeier E, Aurich JC, Govekar C, Hoffmeister HW, Klocke F, Peters J, Rentsch R, Stephenson DJ, Uhlmann E, Weinert K, Wittmann M (2006) Advances in modeling and simulation of grinding processes. CIRP Annals-Manufacturing Technology 55(2):667–696, DOI 10.1016/j.cirp.2006.10.003

Denkena B, Becker JC, Gierse A (2004) Examination of basic separation mechanisms for machining concrete and stone. Production Engineering Research and Development 11(2):23–28

Franca LFP, Mostofi M, Richard T (2015) Interface laws for impregnated diamond tools for a given state of wear. International Journal of Rock Mechanics and Mining Sciences 73:184–193, DOI 10.1016/j.ijrmms.2014.09.010

Herbrandt S, Ligges U, Ferreira MP, Kansteiner M, Biermann D, Tillmann W, Weihs C (2016) Model based optimization of a statistical simulation model for single diamond grinding. Tech. rep., SFB 823 Discussion Paper 11/2016, TU Dortmund University

Huang D, Allen TT, Notz WI, Zeng N (2006) Global optimization of stochastic black-box systems via sequential kriging meta-models. Journal of Global Optimization 34(3):441–466, DOI 10.1007/s10898-005-2454-3

Jones DR, Schonlau M, Welch WJ (1998) Efficient global optimization of expensive black-box functions. Journal of Global Optimization 13(4):455–492, DOI 10.1023/A:1008306431147

Munjiza A, Owen D, Bicanic N (1995) A combined finite-discrete element method in transient dynamics of fracturing solids. Engineering Computations 12(2):145–174, DOI 10.1108/02644409510799532

Picheny V, Wagner T, Ginsbourger D (2013) A benchmark of kriging-based infill criteria for noisy optimization. Structural and Multidisciplinary Optimization 48(3):607–626, DOI 10.1007/s00158-013-0919-4

Raabe N, Rautert C, Ferreira M, Weihs C (2011) Geometrical process modeling and simulation of concrete machining based on delaunay tessellations. In: Ao SI, Douglas C, Grundfest WS, Burgstone J (eds) Proceedings of the World Congress on Engineering and Computer Science 2011 Vol II, WCECS '11, International Association of Engineers, Newswood Limited, Lecture Notes in Engineering and Computer Science, pp 991–996

Raabe N, Thieler AM, Weihs C, Fried R, Rautert C, Biermann D (2012) Modeling material heterogeneity by gaussian random fields for the simulation of inhomogeneous mineral subsoil machining. In: Dini P, Lorenz P (eds) SIMUL 2012: The Fourth International Conference on Advances in System Simulation, pp 97–102

Weihs C, Raabe N, Ferreira M, Rautert C (2014) Statistical process modelling for machining of inhomogeneous mineral subsoil. In: Gaul W, Geyer-Schulz A, Baba Y, Okada A (eds) German-Japanese Interchange of Data Analysis Results, Springer, Cham, CH, pp 253–263, DOI 10.1007/978-3-319-01264-3_22

Zienkiewicz OC, Taylor RL (1977) The Finite Element Method, vol 3. McGraw-hill, London

Semantic Multi-Classifier Systems for the Analysis of Gene Expression Profiles

Ludwig Lausser*, Florian Schmid*, Matthias Platzer, Mikko J. Sillanpää, and Hans A. Kestler**

Abstract The analysis of biomolecular data from high-throughput screens is typically characterized by the high dimensionality of the measured profiles. Development of diagnostic tools for this kind of data, such as gene expression profiles, is often coupled to an interest of users in obtaining interpretable and low-dimensional classification models; as this facilitates the generation of biological hypotheses on possible causes of a categorization. Purely data driven classification models are limited in this regard. These models only allow for interpreting the data in terms of marker combinations, often gene expression levels, and rarely bridge the gap to higher-level explanations such as molecular signaling pathways.

Here, we incorporate into the classification process, additionally to the expression profile data, different data sources that functionally organize these individual gene expression measurements into groups. The members of such

---

Ludwig Lausser · Matthias Platzer · Hans A. Kestler
Leibniz Institute on Aging, Jena, Germany
✉ [ludwig.lausser,matthias.platzer,hans.kestler]@leibniz-fli.de

Mikko J. Sillanpää
University of Oulu, Finland
✉ mikko.sillanpaa@oulu.fi

Ludwig Lausser, Florian Schmid, Hans A. Kestler
Medical Systems Biology, Ulm University, Germany
✉ [ludwig.lausser,florian-1.schmid,hans.kestler]@uni-ulm.de
* contributed equally
** corresponding author

ARCHIVES OF DATA SCIENCE, SERIES A
KIT SCIENTIFIC PUBLISHING
Vol. 1, No. 1, S. 157–176, 2016

DOI 10.5445/KSP/1000058747/09
ISSN 2363-9881

a group of measurements share a common property or characterize a more abstract biological concept. These feature subgroups are then used for the generation of individual classifiers. From the set of these classifiers, subsets are combined to a multi-classifier system. Analysing which individual classifiers, and thus which biological concepts such as pathways or ontology terms, are important for classification, make it possible to generate hypotheses about the distinguishing characteristics of the classes on a functional level.

1 Introduction

The high dimensionality of biomolecular data is one of the major challenges for machine learning algorithms in the field of bioinformatics. The enormous amount of measurements (e.g. gene expression levels) complicates the development of reliable and interpretable models. Initial feature selection can improve the performance of a trained model. This type of model reduction can aid in identifying causes for the predictive ability of the model, which can then further be validated in other experiments. However, feature sets derived in purely data driven or model driven feature selection processes rarely allow a functional interpretation. Measurements are typically selected according to a mathematical performance measure and without respect to known relationships or dependencies. Therefore, these feature sets can rather be regarded as a collection of diverse fragments then as a description of biological processes such as molecular signaling cascades or pathways.

Functional relationships and dependencies can rarely be inferred from a single dataset. Additional knowledge in the form of meta information, i.e. information about information, is needed for grouping or selecting the measurements in an interpretable way. This information can be extracted from a large corpus of biological literature and databases, see e.g. Galperin et al (2015) for an overview of current molecular databases. It aids in focusing on the construction of dedicated feature sets for a single biological process or a small set of biological processes.

The idea of incorporating meta information in the training of predictive models is not new. An overview on recent approaches is given by Porzelius et al (2011). They can mainly be divided into two categories. The first one consists of algorithms that try to guide traditional feature selection processes. For example, Binder and Schumacher (2009) incorporate knowledge on signaling pathways into a boosting model by penalizing the score of the single base learners. Johannes et al (2010) developed a version of recursive feature elimination that is guided by the structure of a protein-protein interaction network. The second category enforces the usage of the given meta information more directly. Abraham et al (2010) construct an intermediate representation of the original measurements. The measurements of one category are replaced by a single feature. Lottaz and Spang (2005) developed an hierarchical classifier system that follows the structure of the gene ontology.

In this work, we propose a knowledge based feature selection algorithm that operates on a predefined vocabulary, i.e. a set of interpretable terms taken form molecular signaling pathways, gene ontology, etc. These verbal phrases are assumed to be reflected in the dataset by a known subset of gene expression measurements. A sparse set of these terms will then be selected and combined in the training of a multi-classifier system.

2 Methods

Classification is the task of predicting the class label $y \in \mathscr{Y}$ of an object on the basis of a vector of measurements, often termed features, $\mathbf{x} = (x^{(1)}, \ldots, x^{(n)})^T \in \mathscr{X} \subseteq \mathbb{R}^n$. The underlying decision criterion is typically formalized as a decision function (a classifier) $c : \mathbb{R}^n \rightarrow \mathscr{Y}$. A classifier $c \in \mathscr{C}$ is initially selected according to a set of m labeled training examples $\mathscr{L} = \{(\mathbf{x}_j, y_j)\}_{j=1}^m$ and denoted by $c_{\mathscr{L}}$ if the chosen training set is relevant:

$$\mathscr{C} \times \mathscr{L} \xrightarrow{\text{train}} \mathscr{C}. \tag{1}$$

An important property of a trained classifier is its risk in misclassifying new, unseen samples

$$\mathscr{R}(c) = \int \mathbb{I}_{[c(\mathbf{x}) \neq y]} dP(x,y). \tag{2}$$

Here $\mathbb{I}_{[]}$ denotes the indicator function.

The risk of a classifier is typically estimated in a resampling experiment as the $r \times f$ cross-validation (Japkowicz and Shah, 2011). Here, the available data $\mathscr{S}$ is split into f folds of approximately equal size. A number of f experiments are performed in which each fold of samples is tested by a classifier trained on the remaining samples. This procedure is repeated for r permutations of $\mathscr{S}$ in order to make the cross-validation error independent from particular data partitions. Let $\mathscr{L}_{ij}$ and $\mathscr{T}_{ij}$ denote the training and test sets of the ith run and the jth split. The error estimation of $r \times f$ cross-validation is then given by

$$R_{r\times f} = \frac{1}{r}\sum_{i=1}^{r}\sum_{j=1}^{f}\frac{1}{|\mathscr{T}_{ij}|}\sum_{(\mathbf{x},y)\in\mathscr{T}_{ij}}\mathbb{I}_{\left[c_{\mathscr{L}_{ij}}(\mathbf{x})\neq y\right]}. \qquad (3)$$

A second important characteristic of a trained classifier is its interpretability. It can be seen as the classifiers ability of giving insights into the properties of a dataset (e.g. identifying important components or dependencies). The interpretability of a trained classifier depends on two distinct properties, syntactical and semantic interpretability.

Syntactical or structural interpretability

The interpretability of a decision function is dependent on its structural properties. The higher the complexity of a decision boundary the lower is its interpretability. The syntactic properties of a classifier can mainly be derived from its concept class $\mathscr{C}$. Possible notions of structural complexity are the number of parameters (Hastie et al, 2001) or the VC-Dimension (Vapnik, 1998).

Semantic interpretability

The interpretability of a classifier is also dependent on the set of measurements that is utilized for a prediction. For instance a selected measurement seems to influence a classification result while a deselected one does not or should not. Other more abstract semantic explanations can be revealed by analyzing the selected feature combinations or structures developed by the trained classifier. Analyses of this type are for example the (gene set) enrichment analysis for the analysis of feature sets (Hung et al, 2012) or principal component analysis (Jolliffe, 2002). The abstract terms that can be detected by these methods are typically strongly affected by noise and should be regarded as fuzzy concepts.

2.1 Feature selection

A common step in the training process of classification models is the selection of informative features (Guyon et al, 2006)

$$\mathscr{C} \times \mathscr{L} \xrightarrow{\text{select}} \mathscr{I} = \{\mathbf{i} \in \mathbb{N}^{\hat{n} \leq n} | i_k < i_{k+1}, 1 \leq i_k \leq n\}. \quad (4)$$

Here $\mathscr{I}$ indicates the set of all sorted and repetition free index vectors of maximal length n. A single element $\mathbf{i} \in \mathscr{I}$ is called a *signature*. It will be denoted by $\mathbf{i} = (i_1, \ldots, i_{\hat{n}(\mathbf{i})})^T$, where $\hat{n}(\mathbf{i}) \leq n$ is the size of $\mathbf{i}$. The elements of a signature indicate the selection of measurements $\mathbf{x}^{(\mathbf{i})} = (x^{(i_1)}, \ldots, x^{(i_{\hat{n}(\mathbf{i})})})^T$ that will be considered in the learning phase of the classifier and for predicting the class label of new unseen samples. It will be called a *feature set* or *feature vector* in the following.

Feature selection is typically a data driven process. That is, a feature set is chosen according to some kind of quality criterion that measures the "informativeness" of the single measurements (univariate feature selection) or a combination thereof (multivariate feature selection). If it can be applied without any knowledge of any other parts of the training algorithm, it can be seen as a preprocessing *filter*.

Feature selection becomes model driven, if knowledge about the concept class $\mathscr{C}$ is incorporated into the selection process. Here, an evaluation criterion is based on the performance (e.g. accuracy) of the classification model $c \in \mathscr{C}$ trained on the current feature combination. The category of model driven feature selection methods comprises the category of *wrappers*, which evaluates general performance measures, and *embedded feature selectors*, which evaluates model specific characteristics.

Data driven and model driven feature selectors share a common search space of $2^n - 1$ feature combinations. It can hardly be analyzed exhaustively due to its exponential growth in n. Most feature selectors are based on heuristic or stochastic search strategies. They usually do not guarantee to find a global optimal solution.

Although data or model driven feature selection clearly reduces the measurements that are involved in generating a decision boundary, it is often questionable if it really simplifies the semantic interpretability of a classifier. Measurements selected according to some kind of performance criterion rarely can be summarized under some interpretable term $\mathbf{v}$. The reason for this is the lack of knowledge about local, temporal or functional dependencies among the measurements. In a purely data or model driven setting, these relationships have to be learned from scratch and often remain undetected.

2.2 Knowledge-based feature selection

In this work we propose a knowledge based feature selection algorithm that allows for incorporating an experimenter's domain knowledge into the feature selection process. A domain expert often possesses knowledge about the experimental setup which typically can not be utilized by the machine learning algorithm. For example, a possible functional grouping of features / measurements is typically known to an experimenter but unknown to the algorithm. The domain expert may also have knowledge about the subject of an investigation, the corresponding measurements and their interactions, etc. For example, an expert in molecular biology has some a-priori knowledge about the molecules that are involved in a certain type of cellular process.

The interactions and relationships described above can typically be summarized by a short verbal phrase that conveys some semantic knowledge to the domain expert (e.g. video-signal, citrate cycle, insulin-secretion). We will call such a phrase an abstract *term* or *word* $\mathbf{v}$. A set of words will be called a vocabulary $\mathscr{V} = \{\mathbf{v}_1, \ldots, \mathbf{v}_{|\mathscr{V}|}\}$. It reflects the external domain knowledge that should be incorporated into an experiment. We will use a word or term $\mathbf{v}$ synonymously with its associated signature $\mathbf{i}$. That is a vocabulary can be seen as a subset $\mathscr{V} \subseteq \mathscr{I}$.

In contrast to a purely model or data driven feature selection, our method constructs feature sets that can be seen as a union of the elements of a subset of the vocabulary $\mathscr{V}$

$$\mathscr{C} \times \mathscr{L} \times \mathscr{V} \xrightarrow{\text{select}} \bigcup_{\mathbf{v} \in \mathscr{V}'} \mathbf{v}, \mathscr{V}' \subseteq \mathscr{V}. \tag{5}$$

That is, the final feature set will include all measurements that are associated to the selected words $\mathscr{V}'$. Without loss of generality, we assume that a typical vocabulary will result in $|\mathscr{V}| \ll |\mathscr{I}|$ and $\forall \mathbf{v} \in \mathscr{V} : \hat{n}(\mathbf{v}) > 1$. In this case a knowledge based feature selection will lead to a reduction of the search space complexity from $2^n - 1$ to $2^{|\mathscr{V}|} - 1$.

Although the final set of features is constructed by selecting a set of words, it is questionable, if the corresponding union of feature sets really reflects the chosen terms. These sets can be overlapping. Their union can implicitly include signatures of additional terms. In order to keep the interpretability of the final signature, we have chosen to couple our knowledge-based feature selection to a multi-classifier system that evaluates each term independently.

2.2.1 Semantic base classifiers (SBC)

Our multi-classifier system is constructed of semantic base classifiers of type

$$c_{\mathbf{v}} : \mathbf{x}^{(\mathbf{v})} \mapsto y. \tag{6}$$

Here $c_{\mathbf{v}}$ denotes a classifier that is restricted to the signature of $\mathbf{v}$ and is therefore associated to this term. The suitability of a term $\mathbf{v}$ is estimated in a 3×3 cross-validation experiment on the learning set $\mathscr{L}$. The signature is therefore evaluated by a multivariate criterion.

A single term $\mathbf{v}^* \in \mathscr{V}$ can be chosen by ranking all terms in $\mathscr{V}$ according to their achieved cross-validation errors

$$\mathbf{v}^* = \arg\min_{\mathbf{v} \in \mathscr{V}} R_{3\times 3}(\mathscr{C}_{\mathbf{v}}, \mathscr{L}), \tag{7}$$

where $\mathscr{C}_{\mathbf{v}}$ denotes a restriction of the chosen concept class $\mathscr{C}$ to the selected term $\mathbf{v}$. The final base classifier $c_{\mathbf{v}^*} \in \mathscr{C}_{\mathbf{v}^*}$ will be trained on all samples in $\mathscr{L}$ and will be seen as an expert in interpreting $\mathbf{v}^*$. In principal, each training algorithm and concept class can be chosen for the underlying training of a semantic base classifier. For our experiments, we have chosen the nearest neighbor classifier (NNC) proposed by Fix and Hodges (1951).

2.2.2 Semantic multi-classifier systems (SMCS)

The multi-classifier system itself can be seen as a decomposable decision rule that is based on an ensemble of semantic base classifiers $\mathscr{E} = \{c_i\}_{i=1}^{|\mathscr{E}|}$, $c_i \in \mathscr{C}$. The final decision rule will be denoted by $h_{\mathscr{E}}$. The training of $h_{\mathscr{E}}$ corresponds to a selection process in which the most suitable set of experts is constructed.

We have chosen an unweighted majority vote h_{maj} as a fusion architecture. It returns the most frequent prediction of the base classifiers as its own prediction and therefore allows a direct interpretation

$$h_{maj}(\mathbf{x}) = \arg\max_{\mathbf{y} \in \mathscr{Y}} |\{c(\mathbf{x}) = y \,|\, c \in \mathscr{E}\}|. \tag{8}$$

The fusion on a symbolic level prohibits interactions on a feature level and conserves the interpretability of the final signature.

The ensemble members are selected in an iterative way. Similar to Equation 7 in each iteration t, a term $\mathbf{v}_t$ is chosen that minimizes the error estimate in a

3×3 cross-validation experiment on the samples of $\mathscr{L}$. The selection of the current term is restricted to those terms that were not selected before. Formally

$$\mathbf{v}_t = \underset{\mathbf{v} \in \mathscr{V}_t}{\arg\min}\, R_{3\times 3}(\mathscr{C}_{\mathbf{v}}, \mathscr{L}), \tag{9}$$

with $\mathscr{V}_t = \mathscr{V}_{t-1} \setminus \{\mathbf{v}_{t-1}\}$ and $\mathscr{V}_1 = \mathscr{V}$. The corresponding base classifiers $c_{\mathbf{v}_t}$ are again trained on all samples in $\mathscr{L}$.

3 Experimental setup and results

3.1 Basic setup

The proposed semantic multi-classifier systems are evaluated in the setting of classifying gene expression profiles. We conduct nested cross-validation experiments to assess their performance (Varma and Simon, 2006) on six different microarray data sets. For the outer cross-validation experiment a 10×10 cross-validation is chosen. The training data of every split is used to select a suitable set of features (signatures) and to train the classifier model. The model selection process for this classifier is based on an internal 3×3 cross-validation as discussed in Sect. 2.2.2. For all experiments, the nearest neighbour classifier (NNC) was chosen as single or base classifier. The semantic classifier systems (SBC and SMCS) were compared to NNCs that use all features and those that incorporate a purely data-driven feature selection process, i.e. the top k features with the highest absolute Pearson correlation to the class label were chosen. The number of features k was predetermined with regard to the chosen vocabulary (k = mean signature size, see Table 2) . All experiments were conducted with the TunePareto-Software for classifier evaluation (Müssel et al, 2012).

Datasets

The experiments are conducted on different two class diagnostic classification tasks. All are related to ageing associated diseases. The data sets are obtained from high-throughput microarray experiments from different technological platforms. All data sets are publicly available from the Gene Expression Omnibus

Table 1 Basic characteristics of the analysed data sets with citation, Gene Expression Omnibus ID (GEOid), feature number (Feat.), sample number (Samp.), and class distribution (Cl.0 and Cl.1).

Dataset	Citation	GEOid	Feat.	Samp.	Cl.0	Cl.1
Alzheimer's disease	Liang et al (2008)	GSE5281	54613	161	74	87
Leukemia	Alcalay et al (2005)	GSE34860	22215	78	21	57
Thyroid cancer	Maenhaut et al (2011)	GSE29265	54613	49	20	29
Lung cancer	Hou et al (2010)	GSE19188	54613	156	65	91
Melanoma	Xu et al (2008)	GSE8401	22215	83	31	52
Pancreatic cancer	Zhang et al (2013)	GSE28735	32321	90	45	45

Table 2 Characteristics of the vocabularies used from the MSigDB (Subramanian et al, 2005) (KEGG, CHROM) and Gene Ontology (Ashburner et al, 2000) (GO), with the number of terms, the number of elements associated to one term (signature) and the total number of covered genes in the database.

	number of terms	minimal signature size	median signature size	mean signature size	maximal signature size	total number of covered genes
KEGG	186	10	53	69	389	5267
GO	3125	10	20	40	492	15992
CHROM	326	5	65.5	91	948	30010

(`http://www.ncbi.nlm.nih.gov/geo/`) database. A brief summary of the data is given in Table 1.

Vocabularies

In our experiments we have used three different sources of meta information:

1. KEGG – Kyoto Encyclopedia of Genes and Genomes (Kanehisa and Goto, 2000) is a collection of molecular signaling pathways,
2. GO – Gene Ontology (Ashburner et al, 2000) is a standardized terminology for the categorization of gene products, here we limited our terms to those that have a set size in the interval from 10 to 500, and
3. CHROM – Chromosomal Location is the position of the corresponding gene within the human genome.

An overview on their key characteristics is given in Table 2. The signatures are extracted from MSigDB (Subramanian et al, 2005) and Gene Ontology (Ashburner et al, 2000). All identifiers have been mapped to gene names. They can be regarded as knowledge of domain experts in molecular biology.

Table 3 Results of the 10×10 fold cross-validation experiments with the KEGG pathways vocabulary. Mean error rates in % $\pm$ standard deviations are given. Feature numbers are given (*All features*), predetermined (*Feature selection*), or averages (*SBC* and *SMCS*).

	Alzheimer's disease (Liang et al, 2008)		Leukaemia (Alcalay et al, 2005)	
	cv-error	features	cv-error	features
All features	9.32 ± 0.72	54613	13.59 ± 0.90	22215
Feature selection	10.31 ± 1.53	69	4.10 ± 0.81	69
SBC (KEGG)	7.37 ± 1.34	325.41	9.23 ± 1.99	174.16
SMCS (KEGG)	7.02 ± 1.02	281.8	8.33 ± 1.51	173.15

3.2 Experimental results

In the following we exemplify our method of semantic multi-classifier systems on selected combinations of vocabularies and data sets. Due to size limitations we do not show all 18 combinations. The selected classification approaches are by no means biased in terms of accuracy, etc., but rather give an arbitrary assignment of data sets and used domain knowledge. In the following each of the tested vocabularies is introduced by a short description first and then validated on two datasets.

3.2.1 Kyoto Encyclopedia of Genes and Genomes (KEGG):

The Kyoto Encyclopedia of Genes and Genomes (Kanehisa and Goto, 2000) is a manually curated collection of molecular signaling and metabolic pathways that regulate different processes in or between cells. A single term from this vocabulary reflects the molecules (more precisely the gene products) that are involved in the signaling process. An example for a KEGG pathway is the *insulin signaling pathway*. It provides the list of molecules that are affected by the binding of hormone insulin to the corresponding receptor of a cell. We tested two datasets using the KEGG-pathways as meta-information. A summary can be found in Table 3.

Alzheimer's disease data set

The Alzheimer dataset was collected by Liang et al (2008) and is available in the Gene Expression Omnibus (GEO) under GSE5281. It comprises brain tissue samples taken post mortem from subjects suffering from Alzheimer's disease (74 samples) and controls (87 samples). Each gene expression profile consists of 54613 probe sets. Applied to all measurements the NNC achieves an cv-error of $9.32\% \pm 0.72$. With feature selection the cv-error is $10.31\% \pm 1.53$. Lower errors are achieved when meta information is used. Coupled to the vocabulary of KEGG pathways a single semantic base classifier achieves an cv-error of $7.37\% \pm 1.34$. A semantic ensemble of three base classifiers achieves an cv-error of $7.02\% \pm 1.02$. Fig. 1a) shows the frequency of the KEGG pathways that are selected in the 10×10 cross-validation. The *insulin signaling pathway* is selected in 91% of the cross-validation splits. It is known that this pathway is impaired in Alzheimer patients (Candeias et al, 2012).

Leukaemia data set

The Leukaemia dataset collected by Alcalay et al (2005) consists of 57 samples of acute myeloid leukaemia with aberrant cytoplasmic localization of nucleophosmin following mutations in the NPM putative nucleolar localization signal and 21 samples without this specific mutation (GSE34860). Each gene expression profile consists of 22215 probe sets. Using all features leads to the lowest performance ($13.59\% \pm 0.90$). With feature selection obtains the best cv-errors ($4.10\% \pm 0.81$). Utilizing the vocabulary of KEGG pathways the best semantic base classifier improves the cv-error (compared to all features) by 4% to $9.23\% \pm 1.99$. The semantic ensemble is able to lower the error rate by another percent ($8.33\% \pm 1.51$). In this case the KEGG pathways *hematopoietic cell lineage* and *cell adhesion molecules cams* are selected most frequently (Fig. 1b). Both terms have been reported in the context of leukaemia (Bonnet and Dick, 1997; Noto et al, 1994).

3.2.2 Gene Ontology (GO):

The Gene Ontology (Ashburner et al, 2000) is currently one of the most prominent attempts of constructing an organized and standardized terminology for

Table 4 Results of the 10 × 10 fold cross-validation experiments with the GO terms vocabulary. Mean error rates in % ± standard deviations are given. Feature numbers are given (*All features*), predetermined (*Feature selection*), or averages (*SBC* and *SMCS*).

	Thyroid cancer (Maenhaut et al, 2011)		Lung cancer (Hou et al, 2010)	
	cv-error	features	cv-error	features
All features	11.22 ± 1.73	54613	8.14 ± 0.61	54613
Feature selection	12.45 ± 2.25	40	6.22 ± 1.51	40
SBC (GO)	11.63 ± 3.34	159.93	4.49 ± 0.74	48.83
SMCS (GO)	6.73 ± 2.16	208.67	4.74 ± 1.14	92.77

the categorization of gene products. It provides an hierarchical ontology of terms that covers three different fields: biological processes, associated cellular components and molecular functions. Most of these terms are linked to manually curated gene lists. The Gene Ontology provides for example the term *cell aging*, which is linked to the list of genes that are known to influence the aging process of cells. The vocabulary of GO terms was tested in two different scenarios (Table 4).

Thyroid cancer

The Thyroid cancer dataset was collected by Maenhaut et al (2011) (GSE29265). Its 49 thyroid samples have been categorised into non-tumour control (20 samples) and thyroid carcinoma (29 samples). The dimensionality of the dataset is 54613. Compared to the experiments with all measurements and data driven feature selection (error rates 11.22% ± 1.73 and 12.45% ± 2.25) the knowledge-based ensemble clearly improves the result. The error rate for the semantic ensemble is 6.73% ± 2.16. A single base classifier is not able to reach this performance (11.63% ± 3.34). Looking at the selected categories in the cross-validation experiment (Fig. 1c) of the ensemble, we find the *chondroitin sulfate metabolic process* term as the one which is most frequently selected (Infanger et al, 2006).

Table 5 Results of the 10×10 fold cross-validation experiments with the chromosomal locations vocabulary. Mean error rates in % ± standard deviations are given. Feature numbers are given (*All features*), predetermined (*Feature selection*), or averages (*SBC* and *SMCS*).

	Melanoma (Xu et al, 2008)		Pancreatic ductal adenocarcinoma (Zhang et al, 2013)	
	cv-error	features	cv-error	features
All features	8.19 ± 1.11	22215	24.67 ± 1.95	32321
Feature selection	8.92 ± 2.39	91	23.56 ± 3.28	91
SBC (CHROM)	7.59 ± 1.14	165.02	22.89 ± 3.24	69.13
SMCS (CHROM)	6.51 ± 1.63	148.39	19.78 ± 3.00	65.79

Lung cancer

The Lung cancer dataset (GSE19188) collected by (Hou et al, 2010) comprises samples of non-small cell lung cancer (91 samples) and adjacent normal tissue (65 samples). Each profile consists of 54613 probe sets. The mean cv-error achieved by the NNC on all features is $8.14\% \pm 0.61$. By using data driven feature selection this result can be improved to $6.22\% \pm 1.51$. On this dataset a single semantic base classifier achieves a slightly better classification performance than the ensemble. The cv-errors are $4.49\% \pm 0.74$ for the base classifiers and $4.74\% \pm 1.14$ for the ensemble. The most frequently selected term is related to ascorbic acid (vitamin C) metabolism (Fig. 1d). Ascorbic acid has been reported to have the ability to kill cancer cells under certain conditions (Chen et al, 2005).

3.2.3 Chromosomal location:

The vocabulary of chromosomal locations (CHROM) can also be used to organize the set of gene expression levels. Here, we restrict ourselves to the human genome. It is organized in 22 pairs of autosome chromosomes and one pair sex chromosomes. Each of the chromosomes can be divided into several cytobands. They can be used to indicate local aberrations. A single term out of this vocabulary gives the index of the chromosome, the chromosome arm (p = short arm, q = long arm), and the cytogenetic bands position on the chromosome arm. For example, *chr17p12* denotes band 1, subband 2 on the short arm of the 17th chromosome. Our experiments with the vocabulary of chromosomal locations are summarized in Table 5.

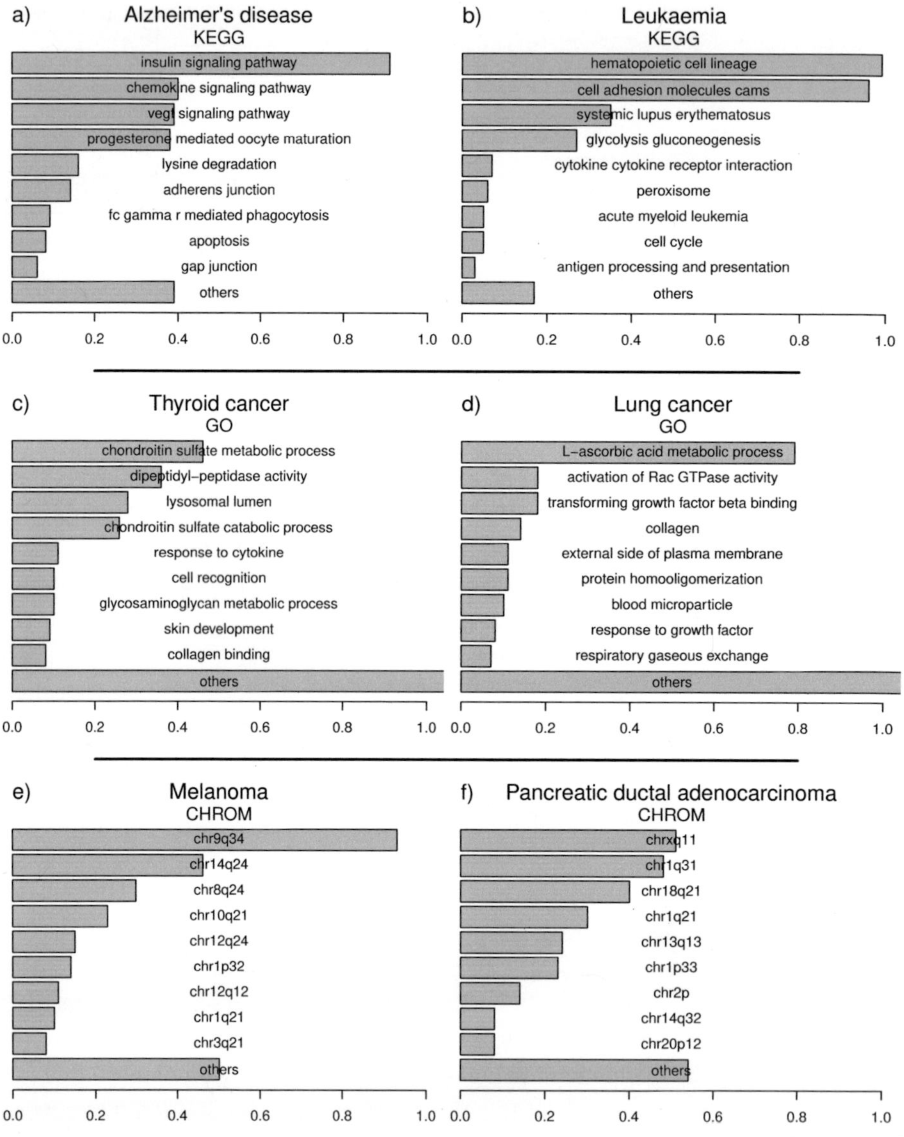

Fig. 1 Frequencies of the terms selected by the semantic multi-classifier system (SMCS) in the 10×10 cross-validation experiments. In total the frequency of 300 terms ($= 10 \times 10 \times 3$) is depicted in each diagram (a to f), normalized to the 100 experiments conducted each. The top nine selected terms are shown. The tenth bar "others" summarizes all categories that are selected less frequent.

Melanoma

The Melanoma dataset (Xu et al, 2008) was collected with the purpose to distinguish between primary melanomas and melanoma metastasis (GSE8401). Both classes are represented by 31 and 52 samples, respectively. The dimensionality of the corresponding gene expression profiles is 22215. For this dataset the data driven feature selection (cv-error: $8.92\% \pm 2.39$) performs worse than using all measurements (cv-error: $8.19\% \pm 1.11$). Using the vocabulary of chromosomal locations (CHROM) as meta information allows to improve the performance. A single semantic base classifier achieves an cv-error of $7.59\% \pm 1.14$. The semantic ensemble improves the cv-error to $6.51\% \pm 1.63$. The most frequently selected chromosomal band is *9q34* (Fig. 1e). It contains the ASS gene which is known to play a role in the cell death in melanomas (Savaraj et al, 2007).

Pancreatic ductal adenocarcinomas

The second dataset tested with chromosomal locations was collected by Zhang et al (2013) (GSE28735). Gene expression values of 45 pancreatic ductal adenocarcinomas and 45 adjacent non-tumour tissues have been measured in profiles of 32321 probe sets. The best results ($19.78\% \pm 3.00$) are achieved by ensembles using the vocabulary of chromosomal locations as meta information. Semantic base classifiers are able to achieve an cv-error of $22.89\% \pm 3.24$. Using all features or data driven feature selection leads to $24.67\% \pm 1.95$ and $23.56\% \pm 3.28$ cv-error, respectively. For this dataset three chromosomal bands are selected with comparable frequencies in the cross-validation experiment (Fig. 1f). The ensemble selects *Xq11*, *1q31* and *18q21* in most of the cases. For *1q31* and *18q21* an association to pancreatic cancer has been reported (Chen et al, 2003; Hahn et al, 1995).

4 Conclusion

We present a knowledge based approach for the design of classifier systems that are interpretable in abstract terms. The basic algorithm incorporates meta information in the form of a vocabulary of signatures (terms) that can be used for constructing a decision rule. The design of the algorithm ensures a high-level interpretability and eliminates the need for revealing an interpretation via reconstruction methods. Our experiments suggest that knowledge based classifiers can be applied beneficially in the field of analyzing gene expression profiles. The constructed models fit into the biomedical context of the analysed diseases. The classification results indicate that selecting only a single term out of a vocabulary neither leads to optimal classification performance nor results in a highly stable selection. Combining a small set of terms improves the classification performance in almost all experiments.

Compared to other approaches the proposed multi-classifier systems excel other approaches by their superior interpretability. Yet, there might be more sophisticated classifier systems that outperform the proposed methods in terms of prediction accuracy. Subsequent work will be focused on the design of classifier systems and other model types that also use continuous outcomes that allow a suitable tradeoff between interpretability and prediction accuracy. For genetic data the presence of close genetic relationships among collected individuals may also bias the results (Habier et al, 2007; Dekkers, 2010), for expression data this is unclear. Integrating meta information in the form of these vocabularies might also be useful for guiding the selection of causal models (Mayo, 1996; Pearl, 2009).

The experiments of this investigation reveal an additional question for the design of a knowledge based classifier system. Although the selected kind of meta information will mainly be determined by the design of a medical/biological study, there may be some a-priori hints on the suitability of a vocabulary of signatures. These hints might be given in the structural properties of a vocabulary (e.g. overlap between signatures) but also in their semantic interpretation (e.g. local information vs functional information). This question can be addressed in more detailed analyses on available sources of meta information for gene expression profiles.

Acknowledgements The research leading to these results has received funding from the European Community's Seventh Framework Programme (FP7/2007–2013) under grant agreement n°602783 (to HAK), the German Research Foundation (DFG, SFB 1074 project Z1 to HAK), and the Federal Ministry of Education and Research (BMBF, Gerontosys II, Forschungskern SyStaR, project ID 0315894A to HAK).

References

Abraham G, Kowalczyk A, Loi S, Haviv I, Zobel J (2010) Prediction of breast cancer prognosis using gene set statistics provides signature stability and biological context. BMC Bioinformatics 11(1):277–291, DOI 10.1186/1471-2105-11-277

Alcalay M, Tiacci E, Bergomas R, Bigerna B, Venturini E, Minardi SP, Meani N, Diverio D, Bernard L, Tizzoni L, Volorio S, Luzi L, Colombo E, Lo Coco F, Mecucci C, Falini B, Pelicci PG (2005) Acute myeloid leukemia bearing cytoplasmic nucleophosmin (NPMc+ AML) shows a distinct gene expression profile characterized by up-regulation of genes involved in stem-cell maintenance. Blood 106(3):899–902, DOI 10.1182/blood-2005-02-0560

Ashburner M, Ball CA, Blake JA, Botstein D, Butler H, Cherry JM, Davis AP, Dolinski K, Dwight SS, Eppig JT, Harris MA, Hill DP, Issel-Tarver L, Kasarskis A, Lewis S, Matese JC, Richardson JE, Ringwald M, Rubin GM, Sherlock G (2000) Gene ontology: tool for the unification of biology. Nature Genetics 25(1):25–29, DOI 710.1038/75556

Binder H, Schumacher M (2009) Incorporating pathway information into boosting estimation of high-dimensional risk prediction models. BMC Bioinformatics 10(1):18–28, DOI 10.1186/1471-2105-10-18

Bonnet D, Dick JE (1997) Human acute myeloid leukemia is organized as a hierarchy that originates from a primitive hematopoietic cell. Nature Medicine 3(7):730–737, DOI 10.1038/nm0797-730

Candeias E, Duarte AI, Carvalho C, Correia SC, Cardoso S, Santos RX, Plácido AI, Perry G, Moreira PI (2012) The impairment of insulin signaling in alzheimer's disease. IUBMB Life 64(12):951–957, DOI 10.1002/iub.1098

Chen Q, Espey MG, Krishna MC, Mitchell JB, Corpe CP, Buettner GR, Shacter E, Levine M (2005) Pharmacologic ascorbic acid concentrations selectively kill cancer cells: Action as a pro-drug to deliver hydrogen peroxide to tissues. Proceedings of the National Academy of Sciences of the United States of America 102(38):13,604–13,609, DOI 10.1073/pnas.0506390102

Chen YJ, Vortmeyer A, Zhuang Z, Huang S, Jensen RT (2003) Loss of heterozygosity of chromosome 1q in gastrinomas: Occurrence and prognostic significance. Cancer Research 63(4):817–823

Dekkers JCM (2010) Use of high-density marker genotyping for genetic improvement of livestock by genomic selection. CAB Reviews: Perspectives in Agriculture, Veterinary Science, Nutrition and Natural Resources 5:1–13

Fix E, Hodges JL (1951) Discriminatory analysis: Nonparametric discrimination: Consistency properties. Tech. Rep. Project 21-49-004, Report Number 4, USAF School of Aviation Medicine, Randolf Field, Texas

Galperin MY, Rigden DJ, Fernández-Suárez XM (2015) The 2015 nucleic acids research database issue and molecular biology database collection. Nucleic Acids Research 43(D1):D1–D5, DOI 10.1093/nar/gku1241

Guyon I, Gunn S, Nikravesh M, Zadeh LA (2006) Feature Extraction: Foundations and Applications. Springer, Heidelberg

Habier D, Fernando RL, Dekkers JCM (2007) The impact of genetic relationship information on genome-assisted breeding values. Genetics 177(4):2389–2397, DOI 10.1534/genetics.107.081190

Hahn SA, Seymour AB, Hoque ATMS, Schutte M, da Costa LT, Redston MS, Caldas C, Weinstein CL, Fischer A, Yeo CJ, Hruban RH, Kern SE (1995) Allelotype of pancreatic adenocarcinoma using xenograft enrichment. Cancer Research 55(20):4670–4675

Hastie T, Tibshirani R, Friedman JH (2001) The Elements of Statistical Learning. Springer, New York

Hou J, Aerts J, den Hamer B, van IJcken W, den Bakker M, Riegman P, van der Leest C, van der Spek P, Foekens JA, Hoogsteden HC, Grosveld F, Philipsen S (2010) Gene expression-based classification of non-small cell lung carcinomas and survival prediction. PLoS ONE 5(4):1–12, DOI 10.1371/journal.pone.0010312

Hung JH, Yang TH, Hu Z, Weng Z, DeLisi C (2012) Gene set enrichment analysis: performance evaluation and usage guidelines. Briefings in Bioinformatics 13(3):281–291, DOI 10.1093/bib/bbr049

Infanger M, Kossmehl P, Shakibaei M, Bauer J, Kossmehl-Zorn S, Cogoli A, Curcio F, Oksche A, Wehland M, Kreutz R, Paul M, Grimm D (2006) Simulated weightlessness changes the cytoskeleton and extracellular matrix proteins in papillary thyroid carcinoma cells. Cell and Tissue Research 324(2):267–277, DOI 10.1007/s00441-005-0142-8

Japkowicz N, Shah M (2011) Evaluating Learning Algorithms: A Classification Perspective. Cambridge University Press, Cambridge

Johannes M, Brase JC, Fröhlich H, Gade S, Gehrmann M, Fälth M, Sültmann H, Beißbarth T (2010) Integration of pathway knowledge into a reweighted recursive feature elimination approach for risk stratification of cancer patients. Bioinformatics 26(17):2136–2144, DOI 10.1093/bioinformatics/btq345

Jolliffe IT (2002) Principal Component Analysis. Springer, Heidelberg

Kanehisa M, Goto S (2000) KEGG: Kyoto encyclopedia of genes and genomes. Nucleic Acids Research 28(1):27–30

Liang WS, Reiman EM, Valla J, Dunckley T, Beach TG, Grover A, Niedzielko TL, Schneider LE, Mastroeni D, Caselli R, Kukull W, Morris JC, Hulette CM, Schmechel D, Rogers J, Stephan DA (2008) Alzheimer's disease is associated with reduced expression of energy metabolism genes in posterior cingulate neurons. Proceedings of the National Academy of Sciences 105(11):4441–4446, DOI 10.1073/pnas.0709259105

Lottaz C, Spang R (2005) Molecular decomposition of complex clinical phenotypes using biologically structured analysis of microarray data. Bioinformatics 21(9):1971–1978, DOI 10.1093/bioinformatics/bti292

Maenhaut C, Detours V, Dom G, Handkiewicz-Junak D, Oczko-Wojciechowska M, Jarzab B (2011) Gene expression profiles for radiation-induced thyroid cancer. Clinical Oncology 23(4):282–288, DOI 10.1016/j.clon.2011.01.509

Mayo DG (1996) Error and the Growth of Experimental Knowledge. University of Chicago Press, Chicago

Müssel C, Lausser L, Maucher M, Kestler H (2012) Multi-objective parameter selection for classifiers. Journal of Statistical Software 46(1):1–27, DOI 10.18637/jss.v046.i05

Noto RD, Schiavone EM, Ferrara F, Manzo C, Pardo CL, Vecchio LD (1994) All-trans retinoic acid promotes a differential regulation of adhesion molecules on acute myeloid leukaemia blast cells. British Journal of Haematology 88(2):247–255, DOI 10.1111/j.1365-2141.1994.tb05014.x

Pearl J (2009) Causality: Models, Reasoning, and Inference. 2nd ed. Cambridge University Press, New York

Porzelius C, Johannes M, Binder H, Beißbarth T (2011) Leveraging external knowledge on molecular interactions in classification methods for risk prediction of patients. Biometrical Journal 53(2):190–201, DOI 10.1002/bimj.201000155

Savaraj N, Wu C, Kuo MT, You M, Wangpaichitr M, Robles C, Spector S, Feun L (2007) The relationship of arginine deprivation, argininosuccinate synthetase and cell death in melanoma. Drug Target Insights 2:119–128, DOI 10.4137/DTI.S0

Subramanian A, Tamayo P, Mootha VK, Mukherjee S, Ebert BL, Gillette MA, Paulovich A, Pomeroy SL, Golub TR, Lander ES, Mesirov JP (2005) Gene set enrichment analysis: A knowledge-based approach for interpreting genome-wide expression profiles. Proceedings of the National Academy of Sciences 102(43):15,545–15,550, DOI 10.1073/pnas.0506580102

Vapnik V (1998) Statistical Learning Theory. Wiley, New York

Varma S, Simon R (2006) Bias in error estimation when using cross-validation for model selection. BMC Bioinformatics 7(1):91–98, DOI 10.1186/1471-2105-7-91

Xu L, Shen SS, Hoshida Y, Subramanian A, Ross K, Brunet JP, Wagner SN, Ramaswamy S, Mesirov JP, Hynes RO (2008) Gene expression changes in an animal melanoma model correlate with aggressiveness of human melanoma metastases. Molecular Cancer Research 6(5):760–769, DOI 10.1158/1541-7786.MCR-07-0344

Zhang G, He P, Tan H, Budhu A, Gaedcke J, Ghadimi BM, Ried T, Yfantis HG, Lee DH, Maitra A, Hanna N, Alexander HR, Hussain SP (2013) Integration of metabolomics and transcriptomics revealed a fatty acid network exerting growth inhibitory effects in human pancreatic cancer. Clinical Cancer Research 19(18):4983–4993, DOI 10.1158/1078-0432.CCR-13-0209

Index